The CNC Cookbook

An Introduction to the Creation and Operation of Computer Controlled Mills, Router Tables, Lathes, and More

E Hess

Scited Publications

Figure 46 used with permission from D. Barger.

ISBN 978-0-9821103-0-0

Note

This book is intended to supply useful information about the topic covered. Some of the activities and topics covered are inherently dangerous, such as the use of power tools and equipment, and the assembly and operation of electronic equipment which operates at high voltage and current levels. The reader should consult a qualified individual to provide assistance when engaging in any such potentially dangerous activities that are beyond their expertise and experience. The author disclaims responsibility for any loss, risk, or liability caused to individuals directly or otherwise, from applying the information provided herein.

Any trademarks are property of their respective owners.

Version 1.0.1

Table of Contents

About the book...

Information about computer numerical control (CNC) is available on the Internet and other places. Unfortunately, the information is scattered, and is written at different levels of detail. This book was motivated by a desire to collect the most essential introductory information into a single volume, written in a single voice with a coherent structure. The hope in doing so was to help those new to the subject eliminate the tedious scavenger hunt for relevant information, and help speed them along to success in whatever CNC related project they envision.

1

Introducing CNC

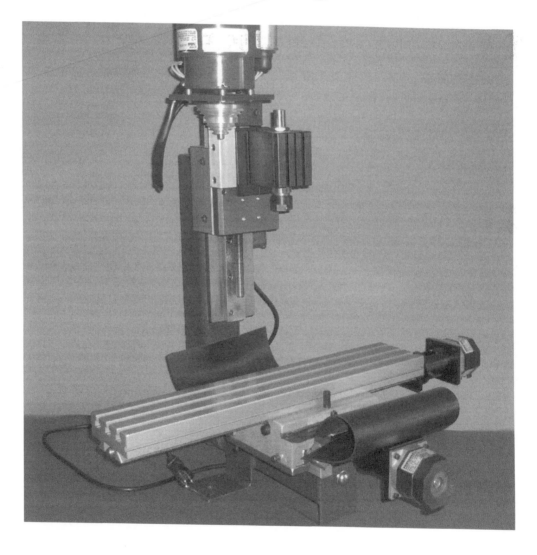

Introduction

This book is intended for those who would like to learn more about CNC technology and apply it to their own projects. CNC stands for Computer Numerical Control, and is a collection of technologies to enable precise computerized control of a variety of machines. If you are interested in creating a CNC controlled device (such as a router table or foam cutting machine) from scratch, or converting an existing CNC (e.g. a bench-top mill or lathe) to CNC operation and don't quite know where to begin, this should be a good resource for you.

Is it possible to for a hobbyist to create a CNC machine from scratch or retrofit an existing machine? Definitely. The concepts aren't outrageously difficult to understand when approached individually, but as a whole there are a lot of different terms and technology to learn about. The goal is to increase your chances of success and reduce the level of stress and time wasted by collecting a lot of the basic information needed together in one place. It is not meant to be the end word on every detail of this expansive subject, but rather as a practical introductory resource for those new to the field.

FAQ

Usually someone starts down the road to exploring CNC when they have a desire to make some high quality part for a personal project or to meet some other need, such as producing a product. The following is a list of some of the big, overarching questions that those first beginning to explore often have:

What is CNC?

What kind of CNC machine do I need?

How do I design a CNC system?

How much work will it require to complete?

What parts are required and how do they work together?

How do I use the machine once I've built it?

Here are some brief (i.e. leaving a lot of stuff out) answers.

What is CNC?

In general, CNC refers to the use of a computer, specialized electronics, and motors to control a machine in a precise and reproducible manner. CNC leverages several different technologies, including software, sophisticated electronics, and specialized mechanical and electromechanical devices. Applications are numerous, and it is used widely in industry for both prototyping (developing a part) and production (producing that part in large quantity), as well as other more mundane operations. New applications and refinements of current approaches are being developed continuously.

Fortunately, these motion control technologies have become much more available to individuals such as hobbyists and craftspeople in recent years. As availability has increased, the range of hobby and small business applications has expanded too. What you can do with CNC technology is limited by your imagination and to some extent by your resources. With creativity you can stretch your resources and create something cool.

What Kind of CNC Machine Do I Need?

You must look inside for the answer to this one; examine your innermost desires, consult your guru, perform an arcane ritual, but by all means think about what you want to use the thing for before you dive in too deep. You may already have a concrete idea of what you want to do. If your answer is something along the lines of "I want a foam cutting machine to make gliders", or "I want to convert my lathe so that I can manufacture custom pool cues" then you have gotten specific enough that you can make steady progress, without too much second guessing. If your answer is "I want to make a machine that will allow me to cut out parts to sell on the Internet", you might want to think a little harder. This is a beginning, but is a bit too vague.

How Do I Design a CNC System?

A big question. This is very application specific. A recommended strategy is to study examples of what others with interests like yours have already done. This topic is covered in some detail in later chapters. There are some common features to creating a CNC system, and these will be covered as you progress through the book. There is no single right answer, however. The possible applications are too diverse.

There are a *bunch* of plans for different types of CNC machines available, which vary significantly in quality. For most common applications, there probably is a set of plans available for something similar to what you want. This is an option for those who want to leave a lot or all of the design decisions up to someone else, and just be involved in building their machine. A drawback of many plans is that they may tie you down to using specific parts that are hard to get (or hard to get at a reasonable price). A goal in the writing of this book was to present the major concepts of designing and building a basic CNC machine or converting an existing machine to CNC operation without tying the reader to an extremely specific course of action.

What Parts, and How Do They Work Together?

There are a lot of different parts, ranging from basic hardware such as screws, bearings, pulleys, couplers, to sophisticated electronic and electromechanical devices such as motor drives, power supply components, and electric motors. There are also many different sub-types of these devices, and they often need to be properly matched to operate correctly. Keep in mind that a CNC machine is a fairly complex integrated system.

Much of the material in this book catalogs these many different parts, and how they do and don't integrate with other related items. Some of this material is encyclopedic and a little dry to read in a straightforward fashion, so it may be best to read the introductions to these chapters to gain a general idea about the subject, and then use the more specific material as needed.

How Much Work is it to Build One?

This depends on how much knowledge you already bring to the table, what tools you have available, and how much you choose to do yourself. The most labor intensive route is to build everything, from assembling the control electronics to the machine itself. This is entirely possible if you are persistent (I've gone this route), and may be the only way to get the machine you want at a price you can handle. You definitely earn it, though.

A less labor intensive way involves buying more of the components of the system already assembled. Buying a conversion kit instead of designing and making your own will also save time, but set you back some cash. Purchasing a ready-made motor controller box is a lot quicker and easier than building your own from parts. You may be able to purchase pre-built functionality for some or many elements of your machine if so desired, trading dollars for time and effort.

How Do I Use it?

Much of the success in operating a CNC machine requires developing some skill with several different pieces of software. The good news is that there are a lot of options available, and some pretty good free ones are available to get you started. A description of these programs and how they relate to each other is discussed in Chapter 14. Additional important topics, such as G-code (the lingua franca of CNC) and practical considerations of operating a CNC machine are covered in Chapters 15 and 16.

The Workflow in CNC

There is a common workflow for design and manufacture of a part using CNC. This is a description that leaves out some (OK, many) details for the sake of clarity:

1. A part is designed in a CAD (Computer Aided Design) program.
2. The part is loaded into a CAM (Computer Aided Machining) program and translated into instructions called G-code.
3. The G-code is checked for correctness and fixed if necessary.
4. The G-code instructions are loaded into a CNC controller program which sends signals to electronic circuitry which control the motors on the CNC machine.
5. The CNC machine cuts the part from a block of material.

If everything is properly set up (material held securely, machine start position set up correctly, and so forth) a part will be created that will closely resemble the designed part. Note, however, that the process of creating G-code instructions to cut a part may be done by hand (i.e. by directly programming it), without the use of CAD and CAM programs. For parts of significant complexity the use of CAD and CAM programs becomes more of a requirement.

Some of the Pitfalls of CNC

Although CNC can be a great boon for people who like to design and build things, it does have some drawbacks which any budding CNC'er should be aware of. CNC machines aren't magic boxes which produce anything and everything without effort. A colleague of mine likes to say that "they aren't a Santa Claus machine." A CNC machine is still a tool, and as with most tools it still requires effort to create something good.

They require practice to use well, so expect to spend some amount of time learning different programs and techniques before you begin making masterworks. As mentioned before, CNC uses a bunch of different software applications, pieces of hardware, electronic devices, and so forth. The relatively large number and variety of things that are integrated to make a CNC machine means that significant energy will be spent up front to get the machine going, and a steady learning process will occur after it is working to create increasingly sophisticated parts.

CNC machines also tend to be application specific, and a given machine will have a limited range of things they can do well. It is important to do your homework and figure out what you want to do with a CNC device before taking the plunge. It is not entirely uncommon to start constructing or converting a machine before realizing that it isn't going to do what was hoped. This may be part of the learning process, but some research might help avoid taking a wrong path.

Automated machinery can create a lot of noise and mess during operation, so plan ahead on where you are going to use the thing once you've got it. A jeweler who uses a small mill to cut wax for castings will not have to worry too much about finding a place to set up shop, as the amount of noise and waste created will be relatively small during operation, and the tiny footprint of the machine makes it easy to place. A woodworker using a relatively large CNC router table, however, will produce a tremendous amount of dust and noise when operating the table, and will have to make special arrangements to deal with this issue such as finding some dedicated shop or garage space, and setting up a good vacuum system.

The Good Stuff

Hopefully I haven't scared you off. Obviously CNC machines have significant benefits or nobody would waste their time with them. Two of the primary benefits of CNC are relatively high accuracy and reproducibility. Complex parts may be cut to exacting specifications; holes will be spaced properly, dimensions will be correct...so long as things are set up correctly initially.

Furthermore, once the instructions to produce a particular part are perfected, that part may be cut over and over. Parts may be cut repeatedly that would be difficult or impossible to do by other methods, with tight tolerances. This is useful for mass production, or when building something that is composed of identical parts, such as a model airplane wing that is created from many identical or nearly identical ribs.

CNC can also be great for prototyping. Sophisticated parts may be designed in a CAD program and then fabricated relatively quickly and accurately on a CNC machine. The CAD file of a given part may be retrieved and modified readily, which can help to shorten the time it takes to develop the final version of a part.

Also, consider that if you have a boring, repetitive, physical task that needs to be done, then CNC may provide an answer. For instance, if you have to have a metal bar cut into evenly spaced sections over and over, or have wire cut and stripped to a specific length repeatedly, CNC automation may provide a solution. Computers, unlike humans, are good at doing the same thing again and again, and will usually do what you tell them to without complaint. (Some might argue with that last statement, but I've found it largely to be true.)

OK, so are you back in the fold? Good! We'll move forward with an overview of a common bench-top CNC machine.

2

Breaking Down a Typical CNC Machine

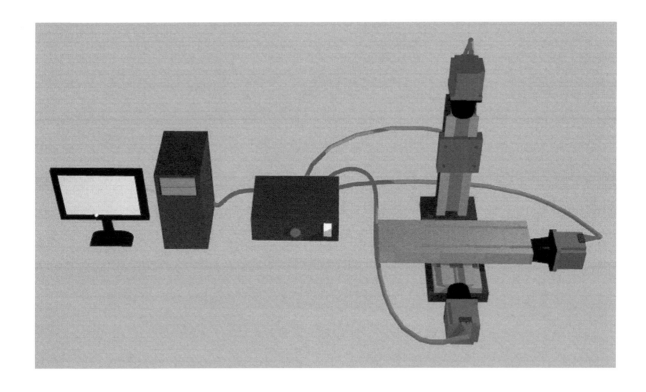

The illustration on the previous page shows the layout of a fairly basic bench level CNC setup. From left to right: a computer runs controller software, which outputs signals to a controller box. The controller box energizes the motors attached to the mill causing them to move the three axes of the mill in a precise fashion.

The following is a photograph of a commercial CNC mill with a cut area of approximately 30 inches by 12 inches.

Figure 1: *A commercial CNC mill manufactured by Haas Automation Inc. Bigger, faster, and stronger than the hobbyist machine, but in many ways, very much the same.*

Obvious differences between this machine and a typical bench-top CNC machine are that the commercial machine is integrated into one unit, and has more features available than the typical hobby level machine does, such as spindle speed control and an automatic tool changer. It can cut larger pieces of material and do it considerably faster. And you'll need a truck to move it.

Despite these differences in size and capacity/capability, the structure of the machines and body of knowledge required to operate and understand what is going on with each is similar. They both have three independent axes, use screws driven by motors to generate motion, and use specialized electronics and a computer running a controller program to control the motor motion. The displays and many of the available functions are similar. The computer instructions (in the form of G-code) are largely the same for operating each machine, and can be produced using the same types of CAD and CAM software.

A Generic CNC Machine

Please note that it is possible to have CNC machines that don't follow the outline given here. This is merely for a common device, such as a bench top mill, lathe, or router table that a do it yourselfer might create or convert to CNC operation.

To help simplify things, I've broken down the hardware, electronics, and software of a basic CNC machine:

The Frame

There is a frame to hold the components, rails, motors, etcetera that make up the CNC machine. In the large commercial machine shown in Figure 1, the frame is cast out of metal and then machined to create a massive, stable base for the rest of the parts of the mill. A Taig bench-top mill, for instance, has a frame built using welded square steel tube. Homebrew CNC machines use a variety of materials, including steel tubing, aluminum extrusion, steel pipe, and even plywood and MDF (medium density fiberboard, a composite material similar to particleboard). If you are going to be designing and building your own machine from scratch, a lot of your success will depend on the solid foundation that the frame provides for everything else.

The Axes

It is common to have one or more axes which move independently. Usually this refers to straight line movement, but may also refer to a different form of movement such as angular movement. In most cases of straight line movement, each axis is composed of the following:

- **Linear rails or ways** to ensure straight line travel.
- **A screw or belt** to transfer power from a motor to move a structure along the rail or way.
- **A motor** to precisely drive the screw, belt or other drive mechanism.
- **A motor mount** to securely hold the motor in place.
- **A coupler or pulley system** to connect the motor to the screw or belt.

Additional Hardware

There are additional pieces that are common to CNC machines, such as
- **A spindle,** which is used for holding and spinning the cutting tools used with machine tools.
- **Limit switches,** which are used to sense when a machine has reached the absolute limit of its travel on a particular axis.
- **Home switches,** which provide a fixed starting point for the machine.
- **Wire,** to carry current to the motors and as a conduit for feedback signals from the machine to the electronics.

The Electronics

Relatively sophisticated electronics are required to control the specialized motors in a CNC machine. Commercial CNC machines usually come with most or all of the required electronics integrated into the machine itself, where hobbyist machines typically have the electronics separate from the machine. Usually, the electronics consist of a personal computer and a device called a motor controller. The motor controller is also referred to as the controller, controller box, or even just 'the box'.

The Computer

This is usually a personal computer running specialized software. The computer sends signals to the motor controller box (described below) through the parallel port (a.k.a. the printer port) or possibly some other interface port, like USB.

The Motor Controller (Controller Box)

The motor controller, or controller box, is a device designed to accept control signals from the computer and then send out the electrical current to control the rotation of the motors on the machine. A basic 'box' usually consists of the following:

- **Motor drives,** which are specialized electronic circuits which receive control signals from the computer and output current to control CNC motors. In some cases, these circuits may receive feedback from the motors to achieve this control. Also referred to as a motor driver.
- **A power supply,** which provides electricity at specific voltages to run the motor drives (which in turn run the motors) and other devices.
- **Additional components** such as relays, fuses, and switches. Relays are often included to automate the turning on of various devices on the machine, such as the motor that spins the cutting tool.
- **An E-Stop:** an E-stop is a switch that is used to shut down a machine quickly in case of a malfunction. Commercial machines all have an E-stop button that may be easily located, and that can be hammered by the operator to quickly turn off the motors and spindle if something goes wrong during a run.

The Software

CNC uses several complementary programs. These programs are usually used in sequence to create a part design and then generate instructions from the design to control the CNC machine. The most important of these programs are:

- **CAD software,** which is used to create detailed drawings of a part.
- **CAM software,** which reads the file created using the CAD software and creates instructions (called G-code) to make the part from it.
- **Controller software,** which interprets the G-code and sends signals to the controller box to move the motors in an appropriate fashion to make the designed part.

Other programs sometimes get into the mix, but these three are usually the most essential. Note that not all CNC machines are designed to cut a part. The above sequence may be used to set up any motion control task, not just typical machining operations for cutting a part.

3

An Overview of CNC Hardware

This chapter gives a broad, brief overview the mechanical and electromechanical devices used in CNC machines. It is intended as a framework to aid in understanding the details in the next few chapters, because a broad array of devices has been developed for motion control applications. Trying to understand the details is probably easier after you have gained a general, less detailed understanding.

Among the most important parts are motors, screws, linear rails, and spindles. There are a lot of different flavors of each of these devices. For instance, there are two major types of motors that are used in CNC machines, namely *servo* and *stepper motors*. There are several different sub types for each motor type as well, such as AC servo motors, DC servo motors, unipolar stepper motors, bipolar stepper motors, and so forth.

Screws also come in different types; some include ball bearings, and some have different thread configurations, among other things. The same can be said for spindles. Understanding the differences between these many parts requires some effort. The focus here will be on the more common, readily available parts used in DIY applications, instead of the ultra-high performance parts used in high end industrial machines.

Motors

Motors are an essential component of a CNC machine. As mentioned above, stepper motors and servo motors are the broadest categories of the motor types used in CNC machines. Either steppers or servos may be used for the types of projects discussed in this book, but each has somewhat different operating characteristics. Stepper and servo motors also have different construction and require substantially different electronic circuitry to run them.

Stepper Motors

Stepper motors are designed to move a fraction of a turn (called a *step*) when the wires of the motor are energized in a specific way, which makes them useful for precise motion control. These motors are different than what most people think of when they think of an electric motor. The shaft of a stepper motor doesn't spin freely when turned, but will hold in place even when there is no electricity being put into it. Another difference is that stepper motors come with more than the normally expected two leads; four, five, six, or eight lead stepper motors are available.

Servo Motors
Unlike stepper motors, the shaft of a servo motor can spin freely when turned by hand. These motors require some form of feedback to operate as CNC devices, and there is a variety of different feedback devices for use with servo motors. There are both AC and DC versions of these motors, as well as brushed and brushless versions. If you are not familiar with these terms, they will be covered in greater detail in Chapter 4.

Screws

The screws used in CNC machine can vary greatly in construction, quality, performance and price. For instance, threaded rod from a hardware or home improvement store is pretty common in low-cost, light-duty, hobbyist machines. In contrast, high precision large diameter ball screws are used for commercial quality machines. The two major categories of screws are *lead screws* and *ball screws*.

Lead Screws
The term lead screw is used to refer to a variety of screw types intended for motion control, ranging from threaded rod to high precision formed screws with fancy thread types and highly engineered nuts. A primary difference between lead screws and ball screws relates to the use of ball bearings, which are not present in lead screws.

Ball Screws
As the name implies, these screws incorporate ball bearings in their construction. The ball bearings are housed in a specialized nut (called a ball nut). The screws have threads that appear like rounded channels in which the ball bearings ride. The ball bearing design was created to make long wearing, precise, and highly efficient (low friction) screws.

Linear Slides and Ways

An essential component of most CNC systems is linear travel, highly controlled straight line movement without slop or deviation from "the straight and narrow." Different components and structures have been designed for this.

Precision Rod and Bushings

High precision (straight with a uniform diameter) round rod with matching bushings is used to produce straight line motion. Round rod may be supported at either end, or along the length of the rod. The bushings are made of either a solid material (such as bronze or plastic) or may use ball bearings in their construction to create a low friction sliding surface.

Linear Rail and Bearing Blocks

High precision rails with a rectangular cross section have also been developed. They typically have bearing blocks that contain ball bearings that roll in channels along the length of the rail. Some bearing blocks are solid, and use a low friction plastic material instead of ball bearings. Linear rail comes in a wide variety of sizes (see Figure 14 for a couple examples) and lengths.

Ways

The term way refers to physical structures such as dovetails that are machined into some material to create mated sliding surfaces. Ways tend to be less efficient (produce more friction) than the two other types discussed here, but they can be very strong and rigid, and can stand up to the tremendous forces produced by some machines. They may also provide a less expensive option for manufacture than other types of hardware mentioned. However, in practice they may be difficult for a hobbyist to create.

Spindles

In CNC speak, the word *spindle* refers to the shaft and associated hardware (bearings, housing, etc) used to hold and spin the cutting tool on mills and similar devices. It is also used to refer to the motor that drives this shaft as well, since these components are often manufactured as a single unit. To specify the motor separately, the term 'spindle motor' is used. Chapter 7 gives a breakdown of the different devices used as spindles. Hobbyists tend to use small to medium sized rotary tools such as a handheld Dremel style 'Moto Tool', a laminate trimmer, or a router. More exotic high performance devices are available as well, but at an exotic price.

4

Hardware: Motors

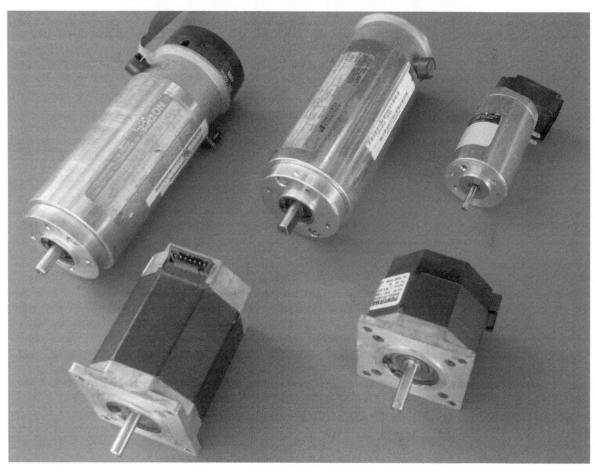

Motors are a big and relatively confusing topic in CNC. As mentioned before, the two major categories of motors used in CNC are servo motors and stepper motors, of which there are a variety of sub-types. This chapter surveys the most common types, as well as their operating characteristics.

Terminology

If these definitions don't make complete sense now, they will become clearer with further exploration. A search in Wikipedia or some other online resource will produce illustrated (and possibly animated) descriptions of different types of motors and how they operate. A decent illustration or animation can show instead of just describe an electric motor's internal parts and workings, and will probably be the quickest route to a clear mental picture of how they function.

Stator: This is the stationary part of the motor which occupies the outer portion (outer cylinder) of the motor, and may contain either permanent magnets or electromagnets.

Rotor: The rotating part of the motor in the center.

Windings: Wire wrapped in a coil which used to create an electromagnetic field. The wire used is magnet wire, which has a thin layer of insulation and can be tightly wound to make a strong electromagnet.

Brushes: Blocks of relatively soft conductive material used with a commutator (see below) to switch the flow of current in a brushed DC motor. These are usually made of carbon, and will wear out over time. Brushes are replaceable in some motors but not in others.

Commutator: A metal structure in a brushed DC motor that acts as a switch when used with brushes. Electrical current is transferred through the brushes to the commutator, and into the winding, generating an electromagnetic field to make the motor spin.

Basic Performance Issues: Motor Speed and Torque

Motors have a property known as torque, which is essentially the twisting or angular force the motor can generate during operation. Consider a wrench being used to twist a nut onto a bolt when changing a tire. The amount of twisting force that is felt at the nut is the torque.

The harder you pull on the end of the wrench, the harder it will twist the nut. Also, the longer the wrench you use the harder it will twist as well, everything else held equal. In general terms, torque is the product of the length of the lever used to produce it (in this case, the length of the wrench), and the amount of force at the end of the lever arm (how hard you pull at the end of the wrench). Therefore, the units of torque are the product of length and force. With an automobile motor, you may already be familiar with the units, which are foot-pounds. For motors used in projects like the ones discussed in this book, the units are typically ounce-inches; still a product of distance and force, but based on smaller units.

Optimal Speed Ranges

Specific types of motors operate with greatest efficiency (and greatest torque), in a particular range of speeds. An analogy is that of a person riding a bike. Most bike riders like to pedal at a comfortable pace (pedal the cranks at a particular rpm). If they encounter a steep hill, they change their gearing so that they maintain roughly the same pace. This doesn't mean that they have to pedal at one particular rate all the time, but that there is a range that they prefer. Falling outside that range will cause the rider to struggle, begin to stall, overheat, and then burst into flames.

OK, I'm kidding about the spontaneous combustion (at least for a cyclist), but for the most part motors behave in a similar fashion. They have a range of speeds in which they best operate. When designing a CNC machine, it is important to consider the type of motor and its optimal operating ranges. The speed at which CNC motors may spin is a function of several factors, including the type of motor, the quality of the motor drive, and the voltage of the power supply, among others.

Stepper motors have a property called *inductance* (a definition is included in Appendix D) which influences their optimal operating range. High inductance stepper motors provide their greatest torque at slower rotational speeds than low inductance stepper motors.

Servo motors tend to operate better at higher rotational speeds than stepper motors and often need to be geared down to turn screws at reasonable speeds with high torque. They are capable of very smooth operation when tuned properly with their companion motor drives.

Gearing, Torque, and Speed

Motors are often geared to help match their characteristics to the application for which they are being used. A motor may be geared up to increase the speed at which it turns a shaft, or may be geared down to increase torque. When gearing up or down however, you exchange one for the other- you don't get both. The only possible exception might be when the existing gearing forces the motor to operate well outside of its normal operating range, and the new gearing corrects for this.

Figures 2 and 3 show a small servo motor that has been geared down so that the screw it is driving turns at a slower rate, but with greater torque. This setup allows flexibility because different gears may be swapped in and out to change the characteristics of the drive system (to improve torque or speed as needed). An additional benefit is that some of the alignment and connection difficulties with direct driving a screw (such as finding a suitable coupler) can be avoided.

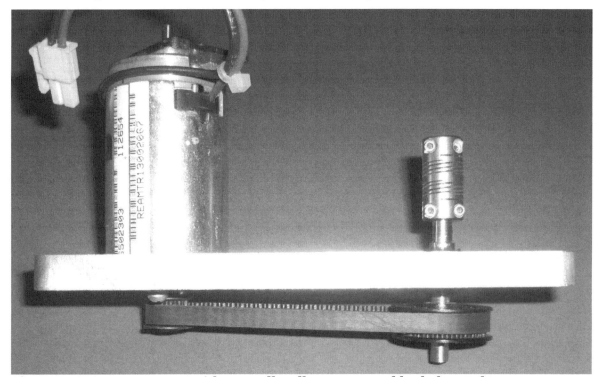

Figure 2: *A servo motor with a small pulley connected by belt to a larger pulley. The coupler (top right) will be connected to a screw and will be driven by the larger pulley.*

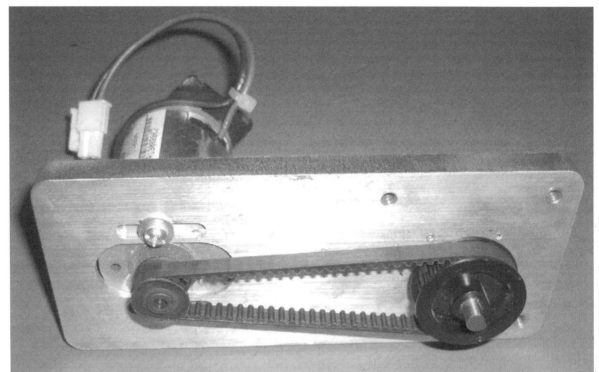

Figure 3: *A different view of the belt, pulley system. Varying the ratio of pulley sizes allows for tailoring of torque/resolution/speed characteristics. In this case torque at the screw is increased while top speed of the screw connected to coupler is decreased. This allows these relatively small motors to drive screws that they wouldn't be able to directly. This plate has been designed to allow adjustment of the distance between the two pulleys for installation and tensioning of the belt.*

The relationship between gearing and its impact on torque and speed is relatively straightforward. If the gear attached to the motor is of an equivalent diameter to that attached to the lead screw or ball screw being driven, then there will be no effect on torque or rotational speed, and it will perform as it would if the motor were directly driving the screw. If however, the gear on the motor is half the diameter of the one on the screw being driven, then the speed of rotation will be half of the direct drive value while the torque is doubled. Conversely, if the larger gear is on the motor, torque will be cut in half while rotational speed doubles.

Gears are characterized by the number of teeth they have, so the combination just described might use a 12 tooth gear for the small gear and a 24 tooth gear for the large gear. When connecting two pulleys of different sizes, you should make sure there is enough contact between the belt and the gears. If the difference in size between the pulleys is very large and they are close together, the smaller pulley may not engage enough teeth with the belt for proper operation. This is remedied by using a longer belt (increasing the distance between the gears) so the belt will wrap around the smaller gear more completely, and more teeth on will be engaged during operation. At least one pulley and belt supplier provides an online calculator program to determine if a particular combination of belt and pulleys will have enough teeth engaged for proper operation, and will also calculate how far apart the pulleys will be with a belt of a particular length.

Voltage and Motor Speed

If you've ever played with an electric motor, you probably know that changing the voltage applied to the motor leads will change how fast it spins, with higher voltages resulting in faster rotation. This is true for both stepper and servo motors. Motors are rated as to their maximum operational voltage, and surpassing the rated voltage may result in motor failure.

Stepper motors are often run at voltages above their rated voltage to achieve higher speeds, but some method of limiting the amount of current is necessary to keep the motors from being damaged when doing this. Newer stepper motor drive designs usually have this current limiting ability built in to them. High voltage operation with older stepper motor drives (ones lacking current limiting) was accomplished by adding resistors (called *ballast resistors*) to limit the current draw at high voltages.

The maximum voltage that your power supply can produce is also a limiting factor on the top speed that your motors may achieve. If you use a power supply that produces only 24 volts but your motors can operate at 35 volts safely, the maximum speed will be less than possible.

Motor Operation, Motor Loading, and Heat

Motors seem somewhat mysterious given that the internal structures (magnets, rotor, stator, coils and so forth) are hidden. Unless you've spent some time tearing motors apart, they are pretty much a black box which you hook up to a source of electricity and watch spin. With a car, for instance, you don't need to know absolutely everything about one to use it effectively, but you do need to understand some basics to keep it running well. This is also true with electric motors. You don't need to know a ton about them to use them successfully, but a little basic knowledge will help a lot.

Motors are a type of *transducer*. A transducer takes one form of energy and converts it into another. A loudspeaker, for instance, converts electrical energy to sound. With motors, electrical energy is converted into angular motion and torque. This conversion process is never one-hundred percent efficient, though. When energy is being put into a transducer and it is not creating the type of output energy it is intended to create for whatever reason, the difference is made up in some other form of energy. With certain types of motors this is primarily heat. It's possible for *a lot* of heat to be generated. As you probably know already, too much heat may degrade or destroy a motor.

For instance, a *brushed DC motor* is a common type of motor used for servos, power tools, and many other devices. If one is under load (encountering resistance to its shaft spinning) it will produce some heat. The greater the load (resistance) it encounters, the more current it will draw to respond to the load, and the greater the amount of heat produced. A real-world example might be a wood router making a shallow cut through a piece of wood at a slow feed rate; this is an 'easy load' for the router motor. As deeper or faster cuts are made, a heavier load is presented to the motor, which results in slower rotational speeds and rising heat production.

From a design perspective, you should use motors that will not always be operating near their limits. This will keep a CNC machine from stalling or losing steps, and will reduce the possibility of overheating a motor. If you discover that the motors you are using will be running near their limits a lot of the time, you may need to make some change to your machine or the way you operate it to properly address this issue.

There are a few options; either use motors that have a higher torque rating, use a screw with a shorter thread pitch (i.e. the threads are closer together), put gearing between the screw and the motor, or use a combination of these things. Additionally, you can run at a lower performance level (e.g. slower speeds) to deal with this issue in some circumstances.

The above comments about motors, load, and heat also apply to the motor used with your spindle. Heavy loads on the spindle must be accounted for by either sufficient motor torque or by reducing the load felt by the motor through other means, such as a slower feed rate or a shallower cut depth.

Characteristics of Stepper Motors

Stepper motors are designed so that they move a fraction of a rotation (a 'step') when the wires of the motor are energized in a particular way. A full revolution is composed of a number of equivalent steps. This number will vary depending on the design of the motor. The ones used for CNC purposes make relatively small steps to provide accurate positioning, and often there are 200 or 400 steps for a single revolution of the shaft. Stepper motors meant for other purposes may have far fewer steps to make a full rotation.

Continuous rotation in a stepper motor is created by repeatedly energizing (running current through) the wire coils in the motor in a particular sequence over and over. You can think of it a bit like how the spark plugs in an automobile engine must fire in a particular sequence to keep the motor running. In contrast to servo motors, stepper motors will hold in a particular spot (i.e. the shaft will not move) even when there is no current flowing through the windings. This is known as *detent torque*.

Also in contrast to servo motors, stepper motors will overheat under somewhat different circumstance than servo motors, such as when configured to spin at faster than normal speeds for extended periods of time. Heavy loads will not cause increased current draw with stepper motors, so overheating doesn't occur due to stalling.

Benefits and Drawbacks

Drawbacks of stepper motors include stalling and cogging (a somewhat jerky motion inherent to the design of a stepper motor) during operation. Stepper motors show resonance in some situations, i.e. they will exhibit low power and rough movement in a certain range of rotational speeds.

Benefits include wide availability; relatively low cost and simpler setup compared to servo motors. A big benefit is that stepper motors do not require feedback to operate like servos do; this is known as *open loop* operation. This allows for simpler and often less expensive setup than with servos. They also tend to be safer to operate than servo motors in certain failure situations. As mentioned before, if a stepper motor stalls for some reason, it will not increase its current draw and start to overheat like many servo motors will. A servo can quickly overheat if there isn't some mechanism to limit the current draw in these situations.

Figure 4: *Three stepper motors. The two on the left are double and single stack NEMA 23 motors respectively, and the one on the right is a NEMA 17 motor with a built in lead screw. Some motors require a special (and sometimes hard to find) connector to connect wire leads to the motor like the one in the middle. The others already have wires permanently attached.*

Step Angle and Steps per Revolution

The terms *step angle* and *steps per revolution* provide essentially the same information about the resolution of a stepper motor. If you were to affix a clock hand to the shaft of the stepper motor, then the step angle is the change in angle of the clock hand produced by a single step. Steps per revolution refers to the number of motor steps required to move the clock hand all the way around once.

If a motor was specified as 400 steps per revolution, the step angle would be .9 degrees, since there are 360 degrees in a full revolution. A 200 step per revolution motor would have a 1.8 degree step angle, and so forth.

Motor Sizes

Standards have been created for stepper motors by NEMA (the National Electrical Manufacturers Association, an industry group). These standards specify things like the spacing of bolt hole patterns and overall case dimensions. When a motor mount is described as "for NEMA 23 motors" it means that any motor built to NEMA 23 specifications should work properly with it; the bolt holes should match up exactly regardless of who made the motor. NEMA 17, 23, 34, and 42 represent progressively larger motor sizes.

Motor Types

There are two common variants of stepper motors; unipolar and bipolar. There are other less common types (e.g. three and five phase steppers) which will not be covered here. Unipolar and bipolar motors require somewhat different motor drives, so it is important to understand the difference before making a purchase.

Unipolar Stepper Motors

The difference between unipolar motors and bipolar motors relates to the internal wiring of the motors. Unipolar motors have five or six wires. Four of the wires are the ends of the motor windings, and the fifth (and sixth in a six wire motor) are connected at the midway point of each winding. The availability of this common wire (also known as a *center tap*) allows for the use of simpler circuitry in the motor drives used with unipolar motors.

It is possible to use unipolar motors with a bipolar motor drive by only connecting four of the leads. However, all leads are required when using a unipolar motor with a unipolar motor drive.

Bipolar Stepper Motors

Unlike unipolar motors, bipolar stepper motors do not have center taps. They require somewhat more complicated motor drive circuitry to drive them, because their design requires that current flows in both directions through a winding, whereas in a unipolar motor the flow of current through a winding is in one direction only.

Motors with four wires are strictly bipolar motors, and may not be run with unipolar drives, unlike six and eight wire motors which may be configured to run with either drive type.

Eight Wire Motors

In terms of wiring, eight wire stepper motors are very flexible and may be configured in many different ways. They may be wired for bipolar operation in series, bipolar operation in parallel, bipolar operation using half of the available coils, and as a unipolar motor by creating externally wired center-taps. See Appendix E for a diagram of how these different configurations of an eight wire motor are hooked-up.

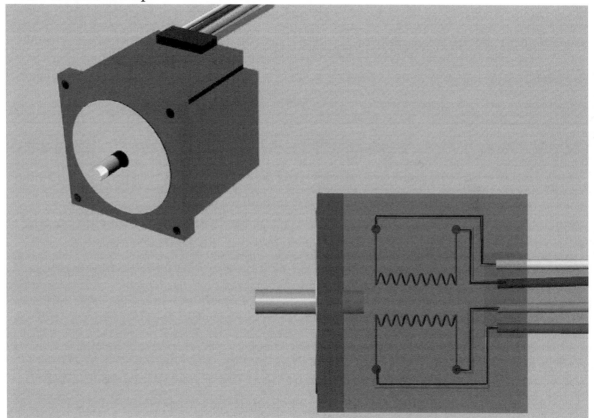

Figure 5: *A stepper motor has internal windings that are energized in a certain sequence to generate steps. This model indicates how two of the wires in a four wire stepper motor are connected to one winding, and the other two are connected to the other. The coils don't actually look like this on the interior of the motor; this is just a conceptual drawing.*

Ratings for Stepper Motors

Stepper motors are also specified as to their 'holding torque.' Holding torque represents the amount of force required to get the shaft of the motor to turn when it is stopped (i.e. 'holding') at a particular position, and is being energized by the motor drive. Stepper motors also have a 'running torque'; the torque they achieve at various speeds during operation. This torque profile is dependent upon various factors such as the quality of the motor drive being used, the maximum current capacity of the motor drive, and the way that the motor is configured (such as in series, parallel, or using a half winding). As a general rule, though, torque decreases as rotational speed increases.

Motors are also specified in terms of their electrical properties, such as current, voltage, and resistance. If you know two of these values and would like to know the third, this may easily be determined by the use of Ohm's law. Ohm's law states that $V = I * R$ where V = voltage, I = current, and R = resistance in Volts, Amps, and Ohms respectively. In this case inductance is substituted for resistance.

Servo Motors

The selection of available servo motors is even more confusing than with that of stepper motors. Servo motors may be either brushed or brushless (discussed below) whereas all stepper motors are brushless motors. The brushed and brushless types require different motor drive electronics and use different feedback mechanisms. There are also AC and DC servo motors.

If this all sounds a little overwhelming, there is a saving grace. In general, DIY and hobbyist servo systems focus on one type, namely the brushed DC servo motor. It is possible to use more exotic servo motor systems, but, brushed DC motors are relatively available, and there are good, reasonably priced motor drives available too. Brushed DC servo motors and their controller electronics will be covered here to the exclusion of the other types of servo motors.

Feedback

The feedback required by servo motors is produced in different ways. For instance, the small servo motors used in radio controlled planes and cars use a potentiometer (a type of variable resistor) as a source of feedback for the controller electronics. A change in resistance indicates how far the motor shaft has rotated.   The feedback devices used for the servo motor systems described in this book are called *encoders*. Used with a motor drive, the encoder provides the feedback necessary to determine how much the motor has rotated and in what direction. Some encoders have been developed that are capable of detecting extremely small movements, allowing for the creation of very high resolution servo systems.

Differences between Brushed and Brushless Motors

Both brushed and brushless DC motors require repeated internal switching of the direction of electrical current to operate. Brushed motors use conductive material, called brushes, which physically contact a structure on the rotor (called a commutator) to act as the switch. In contrast, a brushless servo motor uses electronic circuitry and special sensors to accomplish this task. The sensors provide information to the electronic circuitry about the position of the motor's rotor, and the circuitry controls current flow in the motor's windings based on this feedback. The motor drives are different for these two types of servo motors. A motor drive for a brushed DC motor will not worked for a brushless motor and conversely, one for a brushless motor will not work with a brushed motor.

Converting a Brushed DC Motor to Servo Operation

It is possible to convert ordinary DC motors not intended specifically as servo motors by addition of an encoder. To do this they must be reversible (the motor shaft must reverse direction when the polarity of the current running them is reversed). However, attaching an encoder to motors not originally intended for servo use can be difficult, so give this careful consideration before purchasing motors for conversion.

A dual shaft motor, one with a shaft extending out both ends, may be a good candidate for conversion. Through-hole encoders (encoders with a hole in their body to accept a motor shaft) may be purchased from encoder suppliers for a range of different shaft diameters. The motors shown at the right and left in Figure 6 have motor shafts that extend out the back and into the body of the encoders mounted at the back. It is also possible to position an encoder off of the motor (on a belt between the motor and the screw being driven for example). This technique is illustrated in Chapter 11, and was used with the motor in the center in Figure 6.

Servo motors can be quite expensive if purchased new. Some surplus stores periodically carry used or NOS (new old stock) servo motors both with and without encoder. If you find some that meet your requirements, it might be worth picking a set up. If you plan to use the encoder that comes attached to the motor, you should make sure that they are compatible with the motor drive you intend to use them with. For instance, if the motor drive you are planning on using requires quadrature operation, then the encoder must be a quadrature encoder. The encoders must also have an appropriate resolution to operate with the other equipment you are using. It is possible to have either too low a resolution or too high a resolution. Chapter 8 gives more details about encoders.

Figure 6: *Three servo motors. The two motors on the left are identical except that the one on the left has an encoder mounted on it (and has a shaft that extends out the back of the motor and through the body of the encoder). The motor on the right has an encoder mounted similarly to the motor on the left. As might be expected, the smaller motor generates much less power than the others.*

Servo Motor Specifications

Servo motors are specified in terms of *continuous stall torque* and *peak torque*. Peak torque is the maximum torque that the motor is able to generate for a very brief time. Continuous stall torque represents the maximum torque that is available when running the motor for very long periods of time. Continuous stall torque is a fraction of the value of the peak torque. With the large motors shown in Figure 6 above, the peak torque is about 400 ounce inches, and the continuous stall torque is about 55 ounce inches, giving a ratio of about 7 to 1. This is a pretty typical value for this ratio. Running a motor beyond the continuous stall torque will increase the possibility of overheating because the motor is unable to dissipate the heat rapidly enough at these current levels. The further beyond the continuous stall torque the motor is run, the faster it will overheat.

Other specifications include continuous and maximum current draw, which are about 4 and 28 amps respectively for the large motor in Figure 6. Manufacturers also commonly give the rotational speed (in RPM) at a particular voltage under no load, and the maximum terminal voltage (the maximum voltage the motor should be exposed to).

A Simple Test for Servo Motors
It has been reported to the author that a basic test for a servo motor is to power it with a single AA battery and observe if it rotates, and how smoothly. Good quality servo motors will spin at this low voltage (albeit slowly), and will do so smoothly with relatively little cogging (jerkiness). This was tried and worked with the large servo motor shown in Figure 6.

How Much Motor Do I Need?
How powerful a motor to use is an important question to answer, for both cost and performance reasons. The answer varies somewhat depending on whether you are using stepper or servo motors.

Stepper Motors
Short and simple answers to this question go as follows. For small mills such as the Sherline and Taig, many individuals have successfully used stepper motors rated in the range of 120-200 ounce inches (holding torque) for the X and the Y axes, and its common to use a bit more on the Z-axis to help deal with the weight of the spindle. A very rough guideline is that if you have difficulty turning the shaft of a lead screw by hand (turning it between your thumb and index finger), then a smaller stepper motor in the 120 ounce inch range may prove insufficient, and will have occasional stalls and missed steps.

Going significantly above this (400 ounce inches or so) is overkill and may cause problems as these motors are powerful enough to cause parts to snap instead of just resulting in a motor stall when the machine bumps up against a limit. Additionally, a stepper motor that is over-specified may limit top speed relative to a motor that is sufficient to the task, but not overly so.

This is described in the following quote from Mariss Freimanis, a designer of servo and stepper drives, in response to the notion that 'bigger is better'.

> *People seem to be fixated on motor holding torque alone. Torque doesn't "get things done", power (as in Watts) does. There are much higher torque (1,200 1,800 and 2,000 in-oz NEMA 34 motors). We have some of each and they are dogs. You buy a big bad-ass 2,000 in-oz motor, give a thought what you give away for the bargain. Start with the premise "there is no free lunch".*
>
> *What you give away is speed. Want lots of holding torque? Believe if some is good, more is better? It comes at the expense of speed. It also comes at the expense of motor smoothness and resonance.*
> *http://www.cnczone.com/forums/showpost.php?p=249645&postcount=181*

This quote is in reference to retrofitting a Bridgeport (read this as 'large, old mill') with large, unnecessarily powerful motors. The same issue is true for smaller machine/motor combinations. It is advisable to aim for motors that are sufficiently powerful, but not massive overkill, so that the motor drives aren't fighting to overcome unnecessarily large detent torque during operation.

Servo Motors

To address the question of 'How powerful should my servos be?' I'll begin with a couple of examples.

The small servo motor shown in Figure 2 ran the axes of a Taig mill fairly well using the gearing shown in Figure 3. These motors have a continuous torque of a few ounce inches, and ran smoothly but also ran hot during operation, so are probably a bit underpowered for this application. To address this problem, the gearing could be changed, or slightly more powerful motors could be used.

The gantry style CNC router table discussed in Chapter 11 uses servo motors that are 55 ounce inches in continuous torque, and 400 ounce inches stall torque (the same ones discussed previously and pictured in Figure 6). These motors are also geared at approximately a two to one ratio, thus doubling the effective torque. Additionally, this machine uses very efficient ball screws with a lead (this term is defined in Chapter 5) of 5 millimeters, or approximately .2 inches. Essentially, this combination of motor, gearing, and screw is overkill here, and smaller motors could be used although in this case there aren't major drawbacks to the surplus power. The machine runs fine (the motors are rarely stressed during operation), and the additional power allows for greater performance during acceleration of the ball screws (spinning them up from low to high RPM quickly).

Another Way of Approaching this Issue

You might ask why I'm giving specific examples such as this and not some plug-n-chug formula. One reason is that there are a lot of variables to consider, and some are rather difficult to accurately assess. Sometimes it is difficult to locate a spec sheet for a motor, for instance, so you have to guess at values for torque. Some motor driver electronics have different performance characteristics, and will perform better or worse in a given setup. Motors don't run independently of other parts of the machine, they are integrated with electronics and connected to mechanical devices which can significantly influence their performance.

Another reason is that there are probably several examples on the Internet which are the same or very similar to what you have in mind, which can give you a direct indication of whether your setup will perform properly. These examples can provide guidance, lower and upper bounds of what will work, and may keep you from going too far off the mark, especially if you're new to this. It makes little sense to go to great lengths to figure out what might work in theory, when you can easily do a little research and determine what actually works in practice.

Should I use Steppers or Servos?

This is a worthwhile question, and one that is easy to make too complex. In an attempt to simplify things, I will avoid a few of the smaller details, and compare the two based on important attributes. Again, I am restricting the category of servo motors to the brushed DC type.

Simplicity: Steppers win this contest hands down. Stepper motors do not require encoders like servo motors do, and don't have the additional wiring from the encoders back to the motor drivers that servos require. They do not require tuning of the motor, motor driver, and encoder feedback system that a servo motor system does. In general, once a stepper motor is correctly wired to its motor driver, it should be good to go. Stepper motors have standardized sizes and bolt holes, where servos come in a greater variety of shapes and sizes, and may require more effort to properly mount to a machine. Probably the biggest problem with steppers is that the initial wiring of six and eight wire motors is a bit complicated, but not terribly so. I could go on, but there is no sense in flogging a dead horse...

Expense: Stepper systems tend to be more economical than servo systems for low power systems such as those used for a bench top mill or lathe. This is in part because of the requirement that servo systems use encoders. It is also easy to find relatively low cost stepper motor drives to use with low power stepper motors, but this isn't true for servos. However, for higher power motors systems used in larger CNC machines, the answer to the cost issue is less clear.

Speed: Steppers tend to perform well at lower speeds than servo motors; i.e. they produce high torque at lower speeds, making them a good choice for direct driving a load at lower RPM. However, torque drops off rapidly at higher speeds. Servo motors have better torque at high speeds than steppers, and they may require gearing down for some applications. Neither type is inherently superior in this regard, one may be better than another depending on the requirements of the application.

Performance: In general, servos are capable of greater acceleration and deceleration with a given load than a stepper motor, and a well set up servo system can perform very smoothly at high speeds. Stepper motors tend to have less smooth operation and can be noisy at higher speeds. These properties depend significantly on the motor drive electronics used with the stepper motor, however.

Safety: Stepper motors tend to be safer to operate for a couple of reasons. When a stepper motor is overloaded, the motor will simply stall. A servo motor, however, draws current based on the load it encounters, and will continue to draw current when stalled, resulting in heat production. This heat can cause damage to the motor if there isn't a mechanism in place to stop the current flow. The second reason is that if something fails in the motor control system for a servo, current may continue to be fed to the motor causing a 'runaway' motor situation, where the motor is stuck in the 'on' position until something dramatic happens. This requires additional safety mechanisms be put in place to deal with this situation. With a stepper system, if something fails, typically the motor just stops.

Size and Efficiency: Servo motors tend to generate more power for a given size than stepper motors. They also tend to be more efficient during operation, drawing substantial current when encountering higher loads.

A Simpler Answer

If you would like an even simpler answer to the servo versus stepper debate, consider the following. For machines with small power requirements (such as a small mill or lathe) steppers will work fine, are simpler to implement, and will probably cost less than a servo system. With systems requiring higher power, such as a decently sized router table, either servo or stepper motor system are reasonable choices. For machines with very high power and performance requirements, a servo system may prove to be a better option than steppers.

This is only a general guideline, and not a rule carved in stone. Both stepper and servo systems have been used successfully in a wide variety of applications, and the virtues of one type of motor may outweigh the virtues of the other type for a particular application.

5

Hardware: Screws

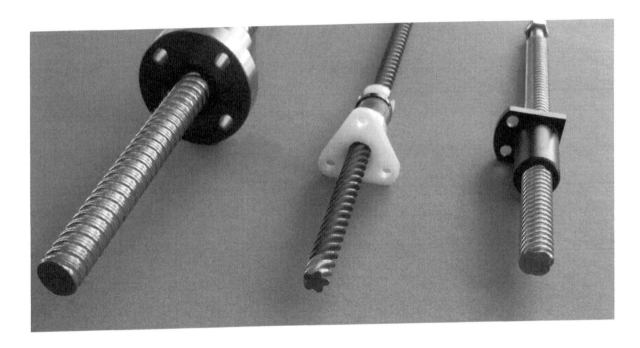

It is common to use screws to move things in a machine tool. They can vary greatly in size, price, and performance. Fortunately, the situation isn't quite as complex as it is with stepper and servo motors. As described earlier, the two big categories for screws are *lead screws* and *ball screws*, which differ in how the threads and nut are constructed.

Terminology

As with the section on motors, I've decided to put some of the more important terminology you will encounter when reading about screws up front:

Thread pitch refers to the distance separating adjacent peaks of the threads on a screw.

Lead describes how far the nut will move in a straight line when the screw is turned one complete revolution.

Efficiency refers to how effectively energy is transferred to the nut when a screw is turned. Low efficiency screw/nut combinations lose a lot of energy because of friction between the screw and nut. High efficiency screws have little friction, and turn motor power into motion more effectively.

Backlash refers to the amount of slop between a nut and screw, and is an important performance characteristic.

Start or starts: The number of *starts* on a screw refers to the number of individual threads on that screw. Most common screws have a single thread, but it is possible to have several different threads on a screw.

More about 'Pitch' and 'Lead'

Pitch or *thread pitch*, is the distance between neighboring peaks on a thread. This is also known as TPI or *threads per inch*, and is equivalent to one divided by the number of peaks per inch of screw. For example, a piece of threaded rod with 20 thread peaks per inch has a pitch of 1/20th of an inch, or .05 inches.

Another important term is the *lead* (rhymes with feed) of a screw, which is essentially how far the nut will travel when the screw is turned one revolution. There is a straightforward relationship between the pitch and lead of a screw. The lead is equivalent to the pitch multiplied by the number of starts in a screw. The formula is:

Lead = (# of starts) * (pitch)

For a screw with 16 threads per inch and a single start, rotating the screw once will move the nut 1/16th of an inch; hence the lead will be 1/16th of an inch. If this same thread had four starts and everything else was held equal, one rotation of the screw will move the nut 1/4th of an inch, so the lead would be 1/4th of an inch.

These values may also be specified in metric which is typical for screws manufactured in Europe and Asia. For instance, a metric ball screw may have a lead of 5 millimeters. The conversion factor is 25.4 millimeters = 1 inch. Regardless of how a screw is specified, controller software can be configured to operate in either English or metric units depending on the preference of the operator.

Lead Screws

Lead screws are commonly a high quality screw (with a matching nut) used for CNC or some other linear motion purpose. A piece of threaded rod and generic nut could be considered a lead screw for a very simple CNC machine, but often this refers to high quality screws specifically manufactured for this purpose. The nuts used with lead screws are usually made out of some fairly hard, slick form of plastic or out of bronze.

Probably the most interesting aspect of lead screws is that they are often manufactured with more than one independent thread on the screw, unlike the single thread design that most people are familiar with. The use of multiple threads offers some unique advantages over single thread screws.

Something Old, Something New...Multiple Threads on a Single Screw

To visualize how a screw can have more than one thread, think of a DNA double helix, the molecule that encodes the chemical blueprint for most life on earth. Each helix winds around the other, and this is like having a screw with two starts. This is illustrated in Figure 7. One advantage of multiple starts is that there is a significant amount contact area between the screw and nut, which helps to eliminate slop during operation. Additionally, they can be designed to have long leads, even up to a few inches per turn. This enables very high speed travel if desired. Single thread screws cannot achieve nearly as long a lead as screws with multiple starts.

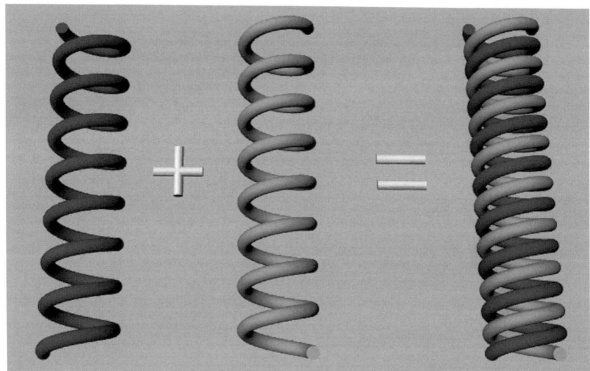

Figure 7: *Two helices (the coils you see here) are joined together to form 'two-start' screw. The helices are offset 180 degrees so that they don't overlap. It is possible to include many more starts than just two by using additional helices, offset from each other by equivalent angles. For instance, a four start screw would have four helices which are offset from each other by 90 degrees. The end on view of the screw in Figure 8 shows how the five starts are oriented relative to each other.*

If you do not know how many starts a particular screw has, you can quickly determine the number of starts quickly by looking at the screw end on, and count the number of individual "bumps". It is surprising how many different threads may be put on a single screw, as shown in Figure 8. Another way to determine the number of starts is to determine how far the nut moves with a single turn of the screw, and then divide by the thread pitch (distance between adjacent peaks). The nut on a four start screw, for instance, should move four times the distance between adjacent peaks when the screw is turned once.

Ball Screws

Ball screws are different from lead screws in that the nut that is used with a ball screw contains ball bearings that contact the thread of a ball screw. These ball bearings continuously recycle themselves (re-circulate) as the nut spins around the screw. The threads of a ball screw are somewhat different in appearance than threaded rod or lead screws; they look like rounded channels for the bearings to roll in. They tend to be more efficient (have less friction between the nut and the screw) than many lead screws, and will require less motor power to move the same load.

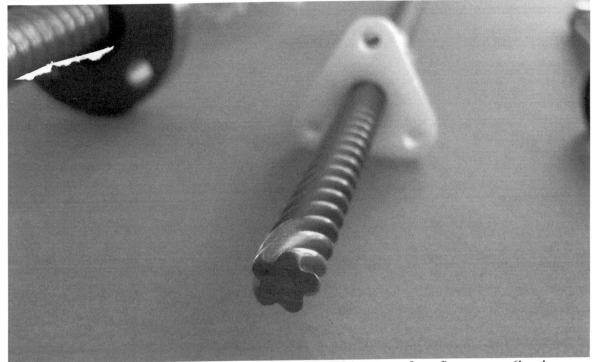

Figure 8: *A lead screw viewed from the end. This screw has five starts (i.e. is composed of five independent threads) which is indicated by the five bumps in the cross section. This results in a fairly steep angle for the five helices that make up the threads.*

Disadvantages of ball screws are that they are more difficult to keep clean in dusty or dirty situation (which affects their durability and performance), and they are prone to *backdriving* (discussed later in this section). They also tend to be pretty expensive.

Rolled and Ground Ball Screws

When shopping around for ball screws, you will probably see the terms *rolled* and *ground*. These terms refer to how a screw is manufactured. The threads of a ground ball screw are created by grinding hardened steel rod, whereas the threads of a rolled screw are created by forming unhardened steel, which is hardened after the process of forming. Grinding is a slower process than rolling and can produce somewhat higher quality screws.

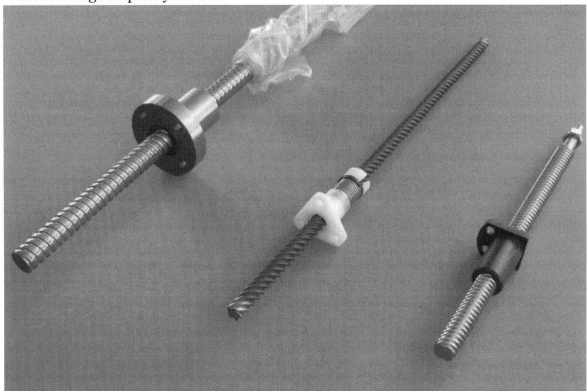

Figure 9: *A ball screw (at left) and two lead screws with their nuts. The ball screw was manufactured by grinding, and the threads act as channels for the ball bearings to roll in. The lead screw in the middle has a split nut with a spring to help eliminate backlash and the one on the right has a nut that is a single piece molded out of plastic. The lead screw on the right has a machined and threaded end for mounting the screw in a bearing or bearing surface. In most circumstances, the ends of screws will be machined prior to use.*

Backdriving

Ball screws, in part because of their high efficiency, may *backdrive* in certain circumstances. What this means is that the ball nut can move down the ball screw of its own accord due the high efficiency (low friction) of the screw, and the steepness of the threads (the angle of the helix). If there is a reasonably heavy load attached to the nut, the downward force will cause the screw to spin and the load will travel downward on its own. This is undesirable because it will require some arrangement to hold the screw in place.

For example, if the Z-axis on a CNC machine has a relatively heavy spindle and a screw that is prone to backdrive, then some significant energy must be continuously put through the motor driving the screw to hold the screw (and in turn, the spindle) in position when the machine is on. Otherwise the screw will start to turn, allowing the spindle to drop down the Z-axis. This generates heat in the motor and may shorten the life of the motor. Additionally, a heavy load supported by a screw that can backdrive may become a hazard if the motor fails, as the load can drop rapidly. Some kind of additional braking mechanism may be required as well, for when the machine is not powered.

Backdriving is more of a concern with large, industrial style CNC machines rather than with the smaller scale machines emphasized here, but it doesn't hurt to be aware of this issue. Just understand that it is possible for some screws which have a vertical orientation (an orientation that is fighting gravity), to begin to turn and allow whatever they are supporting to drop downwards on their own unless they are held in place. Think of how a ball will roll down a slope unless held in position. The greater the weight supported, and the steeper the angle (the more vertical the orientation), the more forceful this effect will be. Some measures will be necessary to hold things in place to stop this unwanted movement from occurring.

If you are able to get your hands on a high efficiency screw with sufficiently steep threads, it is interesting to move it from horizontal to vertical and watch the nut slowly begin to spin on its own, picking up speed as it goes along.

Nuts

Screws need a traveling companion to be useful, namely the nut. The nuts used with lead and ball screws can be relatively simple or highly engineered devices. The primary traits that make a nut a good mate for a screw are the amount of play between the nut and the screw (known as *backlash*), and the relative ease with which the nut is driven by the screw.

Backlash

Backlash refers to the amount of axially oriented (i.e. in the same direction as the length of the screw) "slop" in a lead screw or ball screw and nut. If backlash exists, when the screw reverses direction, the screw will turn a small amount freely before its threads engage and begin to push the nut back the other direction.

Low or zero-backlash is a desirable trait in a screw and nut combination because it helps to ensure precise operation of a CNC machine. It is possible to compensate for backlash in a lead or ball screw in the CNC controller software if necessary, but it is preferable to have a screw/nut combination with little or no backlash to begin with. Significant backlash can result in a machine cutting something other than what is intended. For instance, when trying to move in a circle on a machine with significant backlash, a somewhat oval shape may be produced.

Efficiency

The efficiency of a screw and nut is dependent largely upon the design of and materials used in the screw and nut and the quality of the manufacturing. Manufacturers have exploited different plastics, metals, thread types and so forth to reduce friction between the nut and screw during operation. The use of ball bearings can produce highly efficient screws, but usually at increased expense and complexity.

The reason that high efficiency is desirable is because it lowers the motor power requirements for moving a given load. Less power is required with an efficient screw / nut combination than would be necessary with a less efficient pair to move the same load. Less energy is spent overcoming friction, making it possible in some situations to use smaller (and potentially less expensive) motors than with a less efficient screw.

Pre-load

The term pre-load indicates that something is under force or pressure for the sake of maintaining contact. An example would include a spring loaded nut designed to have zero backlash. These nuts are split into two parts, and employ a spring or other mechanism to force the leads of the nut against the threads of the screw, placing the threads of the nut 'under load' (pressed into contact with the threads) in all circumstances. The center screw in Figure 9 has a split nut with a spring, designed to help eliminate backlash by pushing the two parts of the nut up against the threads of the screw.

The term pre-load may be used in other situations where some device has a load or force applied to it in a particular orientation for the sake of maintaining constant physical contact. For instance, the ball bearings in a ball screw or a bearing block used with a linear rail may be under a certain amount of pressure to force the ball bearings up against the sliding surfaces to help eliminate slop.

Related Hardware

To work well, screws need to be mounted securely, spin easily, and be properly coupled to the motor that drives them. This involves the use of hardware such as bearings, couplers, pulleys, and support structures such as bearing blocks.

Shaft Couplers

To connect the shaft of a motor directly to a lead screw or ball screw, a shaft coupler is used. A shaft coupler is a device designed to join two shafts together, and to allow the connection of shafts with different diameters if necessary. Shaft couplers may be high precision, heavy duty devices or something as simple as a piece of plastic tubing for a very light duty design.

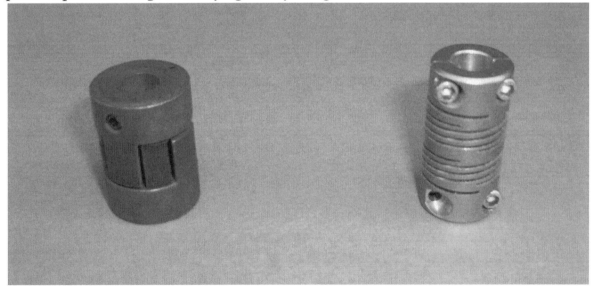

Figure 10: *Two small shaft couplers: these are designed to connect two .25 inch shafts together and both are designed to compensate for some misalignment between shafts.*

Commercial shaft couplers are designed to handle misalignment and different shaft diameter combinations, and it can be difficult to find a suitable one for a given situation (such as when you have an unusual combination of shaft diameters for the motor and screw). In the conversion of the Proxxon mill (Chapter 10), the problem was attacked by creating an adapter that allowed the use of a relatively inexpensive and easily available coupler. Certain types of couplers can be much cheaper to buy or easier to make than others. For instance, a coupler that connect two shafts of identical diameter will probably be easier to make or cheaper to purchase than one that connects a couple of different shaft diameters.

Bearings and Support

Screws require some kind of support at their ends. It is possible to run one end (the end that is not attached to the motor) without support in some circumstances. Typically this is done when the screw being used is relatively short relative to its cross-sectional area. The Taig mini mill uses screws that lack support at one end, but it works fine because the screws are relatively short and stout. They will not flex to any significant degree during operation.

The Forces Encountered at the Bearings

Screws are commonly mounted in bearings at their ends. These may be actual ball bearings or bearing surfaces such as a bushing or even just a hole in some material like metal or plastic. Whatever the type of bearing, it must be able to able to handle the forces from the screw that are created when the machine is operating. The screw is designed to push something back and forth during operation, but when it does this, it itself is pushed back in the opposite direction. To borrow from Newton: *"To every action, there is an equal and opposite reaction."*

When a screw in a CNC machine is being turned by a motor, it generates force to push a gantry (or whatever structure it is connected to) along. In addition, it also must generate force to push the cutting tool into whatever material it is cutting. At the same time, however, an opposing force is transferred down the screw to the bearing surface at its end. The bearing and bearing block must be able to resist this force and remain in place for the CNC machine to operate properly. If the shaft isn't firmly held in place (i.e. it gives a bit when the screw pushes back on it), the machine will lose both precision and accuracy. Figure 11 illustrates this concept.

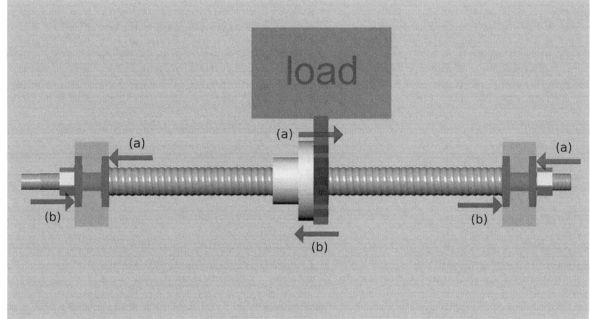

Figure 11: *The bearings at the ends of a screw must resist the forces that are transmitted down the shaft of the screw during operation. The blocks supporting this screw have two bearings at each end. If the screw and nut are pushing a load to the right (the arrow in the center labeled a), the forces felt at the bearings at the end will be in the opposite direction. If the screw/nut is pushing the load to the left (center arrow labeled b), the forces felt at the bearings will be towards the right.*

Also, understand that the forces on the screw are in two directions. The screw will be both pushing and pulling at different times during operation where the screw meets the frame of the machine. The bearings, nuts, and so forth will need to resist movement of the screw in both directions, and should be installed in a way that does so. It is possible to do this at one end or at both ends. With a heavy-duty design, it is probably best that both ends can resist both pulling and pushing forces. Lighter duty designs may not require this much support. As mentioned before, it is even possible to run a screw unsupported at one end, so long as the supported end can resist both pushing and pulling. Screw manufacturers will sometimes provide diagrams of the different ways a screw can be supported (different arrangements of bearings, bearing blocks, and so forth).

Specialty Bearings

Many of the bearings that people are familiar with aren't specifically designed to resist forces pushing on their side to a great degree, they are primarily meant to allow a shaft (or some other structure) to spin. Heavy duty CNC machines employ specially designed bearings, such as thrust bearings or angular contact bearings, which are engineered to withstand these high, axially oriented (i.e. pushing against the side of the bearing) forces. Commonly available bearings like those used for inline skates for example, may be adequate for light weight DIY applications. However, it may be necessary to use these other types of bearings in some heavy duty designs where high forces are being generated during operation.

Shaft Alignment

Shaft alignment is critical. It can be difficult to perfectly align two shafts, such as the motor and screw shafts, exactly. This is especially true if you are making some of the parts yourself, such as motor mounts. There are a few different strategies for dealing with this issue:

- Use a coupler which is designed to accommodate some misalignment, such as shown in Figure 10.
- Use or create motor mounts that allow some adjustability in positioning of the motor shaft relative to the screw shaft.
- Use a pulley and belt system to couple the motor shaft to the screw shaft.

Whipping

It is possible for screws to whip up and down when turned at a particular speed, and this becomes more of a problem with longer lead and ball screws. Many structures exhibit resonances like these, and it is possible to have problems if this issue is not considered. If you would like to see an interesting example of a resonance problem, search for 'Tacoma Narrows Bridge' on the Internet.

The tendency for a screw to whip is a function of screw length, screw diameter, rotational speed, and the way that the screw is fixed at its ends. It may be necessary to make design changes or restrict operating conditions to counteract this effect when using a relatively long screw. Screw manufacturers typically have information about which combinations of rotational speed, length, and screw diameter will produce this, and have published charts which show what the critical speeds are. Ways to fix potential whipping problems include changing the length or diameter of the screw, and changing the manner in which it is held at its ends.

Machining of Screws

Lead screws and ball screws may be purchased with or without machined ends. The machining of screws is a tricky business, requiring skill and good equipment to achieve the necessary levels of concentricity (alignment of the centers of both the screw and the machined surface). If you are going to buy screw material without machined ends, it is a good idea to make sure you have firm plans on how to get the screws properly machined. Purchasing screws with machined ends may appear to be more expensive at first, but when you consider the cost of getting raw screw material machined, they may prove to be a bargain. There are ways to use screws without machined ends, but in general they are a compromise, and tend to be used with small, lightweight machines.

6

Hardware: Slides and Ways

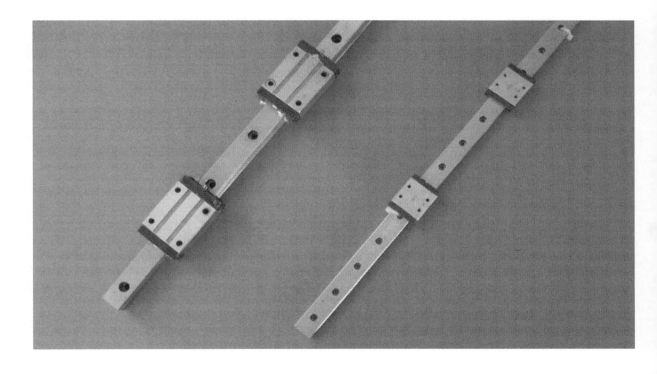

Many devices require some form of linear travel to do what they are supposed to do. Examples include commonplace devices like drawer slides or even a vise. These devices, however, do not take this concept to nearly the extremes that CNC machines do.

In almost all CNC machines, high precision motion is essential, with extremely little slop and deviation from the straight and narrow allowed. There are a lot of different types of linear slides devices available. They all are designed with the same goal in mind, to move things in a straight line.

Common linear slides devices include:

- Round rod and linear bushings
- Linear rails and bearing blocks
- Linear ways
- V-groove rail and bearings

An Aside – Purchasing Linear Slide Devices

Please note that commercial linear slide mechanisms can be quite expensive if purchased retail. It is advisable for a hobbyist trying to construct a machine from scratch to look into surplus and auction sites to reduce costs significantly. It is also possible to improvise and save a large amount of money by using things available at home improvement centers. Some hobby style designs creatively use skate bearings, drawer slides, pipe, and so forth. These applications are usually light duty, but it is quite possible to achieve good results.

Round Rod and Bushings

Precision round rod is commonly used with matching bushings of different styles to make a linear slide mechanism. There are several bushing types including those made out of bronze, plastic, or those that contain ball bearings that re-circulate in a continuous loop. The type of bushing used may depend on the type of machine being constructed. For example, a machine designed to come into contact with food products or medical devices may require plastic bushings which are relatively hygienic, and may be cleaned easily (like some solid plastic ones). OK, so this example probably doesn't apply to the majority of the people reading this book, but maybe someone somewhere needs a high precision automated cake decorating machine.

Brass bushings, such as those shown in the picture that follows, encircle the rod and can only be used with relatively short lengths of rod because the rod can only be supported at its ends when they are used.

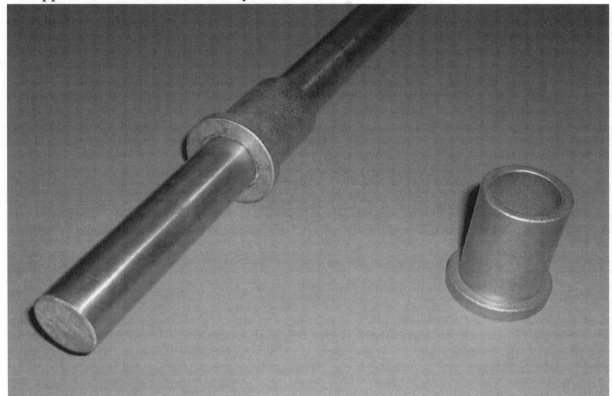

Figure 12: *Precision steel rod and matching brass bushings: Some rod is pre drilled and can be purchased with matching metal supports. The use of supports along the length of the rail requires bearings that are open on one side (they are sort of C–shaped when viewed end on) so as not to interfere with the supports. The brass bushings shown here are closed, so will limit the length of the rail with which they may be used.*

In short lengths, round rod may be used free standing, supported only at the ends. As might be expected, larger diameter rod will deflect (dip or bend) less over a given span than smaller diameter rod. For long lengths of rod, it is essential to support the rails so that they do not deflect significantly from whatever weight they are supporting (see Figure 13).

Figure 13: *A CNC router table under construction. It's a bit difficult to see, but the linear rail for the long axis (the X-axis) is supported almost the entire length to avoid deflection due to load (1). The short unsupported sections are not a problem. The pillow block bearings (2) are open on one side, allowing the use of supports for the rails. There are two pillow block bearings used per side.*

Some CNC router designs incorporate relatively long stretches of unsupported rod, which is not a good design choice as they will deflect downward under load. A situation where this may be OK is when there is very little weight to support and high forces will not be generated during operation. One example is the pick and place machine shown in Figure 37. Short stretches of unsupported rod (as seen in many Z-axis designs) are perfectly fine in some cases, though.

Linear rod may also be used with bearing cartridges and pillow blocks. A pillow block is a metal housing which holds the cylindrical bearing cartridges, and has a flat surface to allow a structure (moving platform, gantry, or whatever is supposed to move along the axis) to be mounted. The bearings re-circulate (similar to the bearings in a ball nut) as the blocks slide along the rods. The ball bearings should be protected from dust as much as possible.

Linear Rails and Bearing Blocks

Linear rails with bearing blocks are another common style of linear slide mechanism. These are essentially square or rectangular rails (they have a square or rectangular cross section) with channels along their length in which the bearings in the bearing blocks contact the rail and re-circulate. These devices are typically high precision and are also relatively easy to install, as they have flat bottoms and have pre-drilled mounting holes frequently spaced along the length of the rail. They can be quite expensive if purchased retail, however. Figure 14 shows a couple of different examples.

A Comment about Scale

As one might expect, the larger a CNC machine is, the more difficult and expensive it is to create. Long precision slides typically cost big money, and designing (for example) a router table that can handle a full sheet (4' by 8') of plywood can be pretty costly. However, if the designer can tolerate lower precision, costs may be lowered substantially. Some of the ways this may be done are described in Chapter 11.

Ways

The term way refers to a linear motion structure that is created from two pieces of material specifically cut to provide mated sliding surfaces. These sliding surfaces must fit closely with a minimal amount of clearance. Many mills incorporate some ways into their construction. Figures 15 and 16 show two different types of way surfaces, a dovetail and box way, used in a Taig mill.

Important Features of Ways

Ways often incorporate a *gib* which is essentially an adjustable surface that may be tightened or loosened to provide optimal performance, so that the ways slide smoothly but have no slop. The gib for the mini mill in the cover photo is barely visible, tucked up against the dovetail. Ways have a relatively large amount of surface contact between the two sliding surfaces. This is both good and bad; the high contact area makes for rugged construction that can withstand significant forces generated by some machines, but also produce a significant amount of friction during operation.

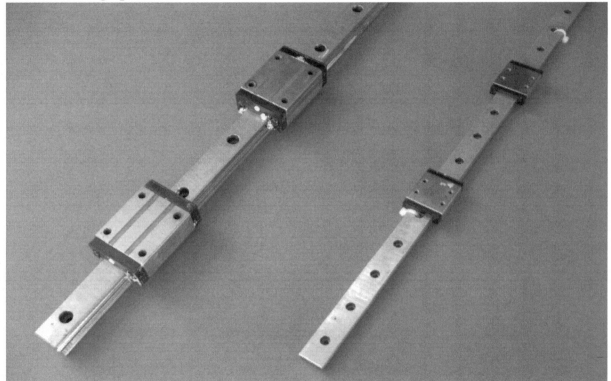

Figure 14: *Two sizes of linear rails with bearing blocks. These rails are easy to mount and have virtually no play; they can be very expensive if not purchased from a surplus or second-hand source, however. The large one on the left, as might be expected, can handle greater loads than the one on the right. Similar to the ball screw in Figure 9, these rails are manufactured outside the US and are specified in metric. Warning: if the bearing blocks come off the rails, then the ball bearings may drop out of the bearing blocks and become lost. It's a good idea to put something in, on, or around the rail to keep the blocks from sliding off the ends before they are installed. Cable ties were used on the rail on the right for this purpose.*

Figure 15 (top) and Figure 16 (bottom): *Box and dovetail ways on a Taig mill. The Y-axis way (not shown) has a slightly different design than either the X or Z axes. Sherline mills use dovetails on all three axes.*

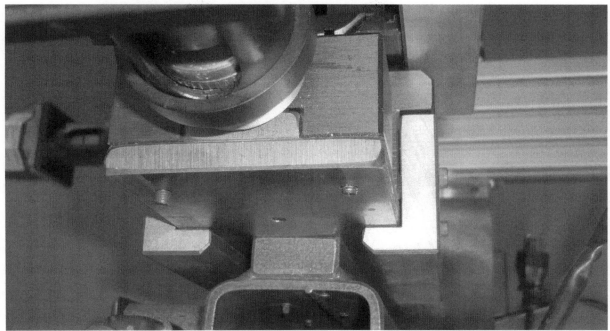

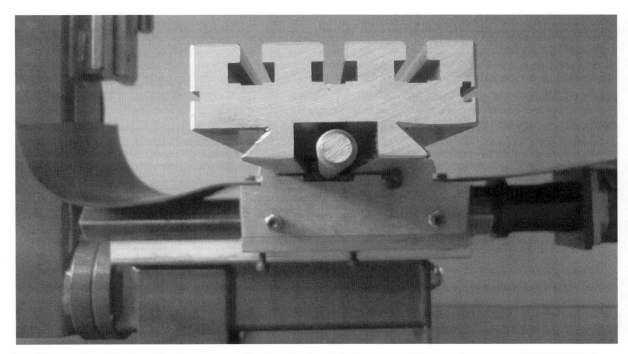

Commercial mills incorporate massive tables with substantial way surfaces to withstand the high forces generated when machining hard materials at relatively high speed, but they must use powerful motors to adequately move these tables. It may be possible for someone to create way surfaces, but it is being mentioned here primarily as a reference to those who are shopping for a mill or lathe to convert to CNC operation.

V-groove Rail and Bearings

V-groove rail and bearings are a relatively flat metal track with a "V" shaped edge (or edges), which are designed to mate with rollers which have a complementary V-shaped indentation. This type of system is usually less expensive than linear rail for a given length and may be suitable for those looking to design a relatively large router table without breaking the bank.

7

Hardware: Spindles

Spindles

In general, the term *spindle* refers to the associated parts that hold and spin cutting tools. From a DIY perspective, this might be a router or rotary tool that holds your drill or router bit. Taig or Sherline mini mills come with a factory built spindle as part of the machine. The motor that drives the spindle is referred to as the *spindle motor*.

On professional machines (like the Haas mill shown in Chapter 2) the spindle is an integrated part of the machine that performs pretty much the same function as on the smaller machine, but may have more features and, obviously, much more power. In some cases the term spindle is applied to the motor, bearings, holding mechanism, and so forth. For example, a router that is being used in a CNC machine would probably be referred to as the spindle because it houses the motor and the other associated parts as a whole. I wouldn't sweat this point too much, however.

Important Considerations

Spindles must be able to handle lateral forces (forces pushing on the side of the cutting tool) because many cutting operations used in mills and router tables generate these kinds of forces. They also must maintain a relatively high level of concentricity (a common 'centeredness' in the rotating parts) because they often spin at very high RPM, and if they are even slightly off they will not function properly. If you are going to be creating a spindle (the bearings and holder excluding the motor) of your own design or use an off the shelf device, this concentricity requirement should factor into your decisions. Unless you have substantial machining skill, you may want to leave spindle creation to a professional manufacturer.

Another thing that may impact a choice of spindle is the relationship between rotational speed and power. Most small cutting devices that hobbyists use in their machines tend to achieve their greatest power at rotational speeds from approximately 24,000 to 34,000 RPM. For some cutting situations these speed may be less than ideal. An example might be a high speed spindle operating in a CNC (wood cutting) router. High rotational speeds may actually result in the cutter burning the wood being cut; the bit can end up being scuffed more than it is cutting, generating a lot of heat. More isn't always better with spindle speeds. Sometimes cutting at slower speeds works better.

Don't let the previous paragraph scare you away from using the devices mentioned. They will probably work perfectly well and many (perhaps most) DIY machines use something along these lines. They are usually the least expensive option available. Having a powerful spindle with adjustable speed may be a bit of a luxury item for a self built machine.

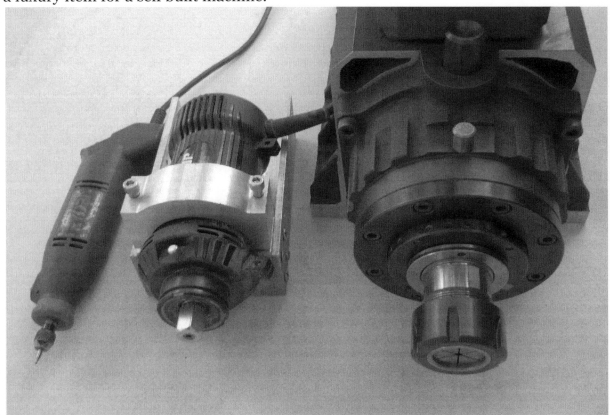

Figure 17: *Three spindles, including a small rotary tool, a laminate trimmer in a holder, and a big, bad 12 HP production grade spindle. More powerful devices allow for faster feed rates but at greater expense, size, and weight. The two smaller devices are typical of devices used in DIY CNC machines. Another common choice is a wood router (not shown here but see Figure 43 for an example).*

If you do incorporate some kind of speed control with these devices (e.g. some external voltage control device) keep in mind that the amount of power generated may drop considerably because they are operating outside of their ideal range. To work well at low rotational speeds, it generally becomes necessary to employ a larger, higher horsepower device. Professional grade milling machines have spindles of ten or twenty horsepower so that they can cut under most circumstances, including making heavy cuts at relatively low speeds.

When considering a given device for use as a spindle, remember that its suitability depends in part on its engineering and in part on the application for which it is being used.

Engraving Spindles

For those interested in engraving applications, such as carving fine patterns and letters onto surfaces, specialized spindles that have a "floating head" are used. What is meant by this is that they have a spindle that is designed to have some 'give' that allows them to cut at a relatively constant and shallow depth to produce good results. When using V-shaped cutters, the depth of cut is especially critical, and irregularities in either the stock being cut or in the way the stock is oriented will result in noticeably poor results. An engraving spindle will move slightly to compensate for these irregularities and maintain a consistent cut depth as the machine carves out letters or a pattern.

8

Electronics: The Computer and the Controller

There are two primary electronic components needed to control the motors and other devices on a CNC machine; the motor controller or controller box, and a computer. Professional equipment commonly integrates the controller electronics, microprocessor, and display into the body of the machine. For most hobbyist or DIY equipment, however, a personal computer and stand-alone controller box make up most of the electronics for automation of a CNC machine.

The Computer

Due to their ubiquity and relatively low cost, personal computers have been a common choice to run CNC controller software and output control signals to a controller box. PCs provide a parallel port (also known as the 'printer port') which can be used to send digital control signals to the electronics in the controller box. Additionally, several good and relatively inexpensive CNC controller programs have become available, making them an attractive choice for the 'brains' of a CNC set up.

The Parallel Port

The parallel port is currently the most common way to send control signals to controller boxes in hobbyist systems. It has some limitations, but for most applications, can be used quite successfully and reliably.

The parallel port provides a large number of input and output lines, and is relatively easy to 'hack'. To utilize the printer port to output signals, it is necessary to understand what the *pinouts* of the printer port are. The term pinout refers to a list and/or diagram of how a connector is wired, and the function of each pin.

A basic breakdown of the available pins in the parallel port is listed below:

Pin or Pins	Pin Type (Usage)
1-9	Output
10-13	Input
14	Output
15	Input
16, 17	Output
18-25	Ground

The output pins are used for sending signals to the motor drives to control motor movement or to trigger relays which control turning the spindle and other devices on and off. Input pins are used to receive signals from devices such as the limit and home switches to determine when a machine has reached a certain location. Specialized and more feature-laden machines will probably use more of these outputs and inputs. For the hobbyist, this will require an understanding of how to interface them properly.

Fortunately there are a variety of products available to ease the chore of accessing and using the parallel port pins. These include breakout boards and relay boards.

Breakout Boards and Relay Boards

A breakout board (or 'breakout') is a device that allows easy access to pins of the printer port. It is possible to directly wire into a parallel port by using a voltmeter to identify which pin goes to which wire and then hardwiring to these pins, but this is confusing and messy. Using a breakout board makes things far easier. There are many different styles available from many different suppliers. They are connected to the printer port via a printer cable or by directly plugging into the port and provide screw terminals or solder holes for connecting wires to the pins of interest. They also allow easy identification of the pins, because the terminals are usually numbered to match the pin to which they are connected.

Breakout boards can range from very simple to complex. Fancier boards may include optoisolation (defined in the Glossary) for protection of the computer, provide a regulated voltage that is at a useful value, and may incorporate relays as part of the board. It is also possible to purchase separate boards for both breakout and relay triggering purposes.

More about Relays

A relay is a switch that may be activated by electrical current. In general, this allows a device that operates at low voltages and currents (such as a personal computer) to turn on and off other, high powered devices. A common CNC usage is to use the PC to turn on and shut off a spindle at the beginning and end of a run. Controller software can be easily programmed to switch several different relays on and off at specific times or after some event. Usually a specific printer port output pin is dedicated to each relay to be switched.

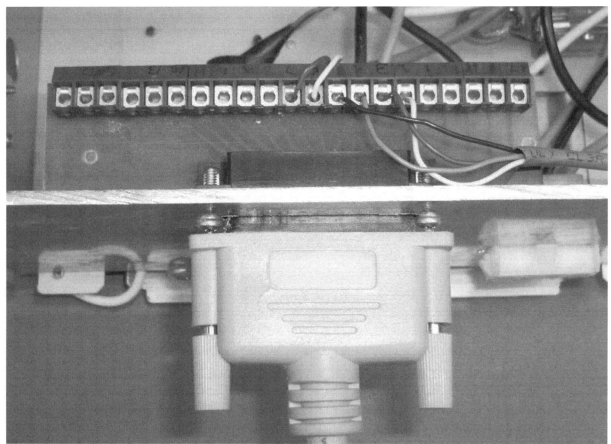

Figure 18: *A breakout board. This simple breakout allows access to the pins of the printer port. A printer cable is plugged in and screw terminals on the breakout allow connection of wires to specific pins for step and direction signals, triggering of relays, inputs for home switches, limit switches and so forth. Commercial breakout boards are available from many sources, and can be basic or include several additional features.*

The Fate of the Parallel Port

The parallel port is being phased out. The USB (Universal Serial Bus) is now being used almost exclusively in new printers, and manufacturers are excluding parallel ports (originally used to communicate with printers) from new motherboard designs because they are no longer needed for printing.

The parallel port is still a perfectly viable choice, however, but it may become necessary to look for a somewhat older computer that comes with a parallel port, or buy a parallel port card. With newer computers, you shouldn't assume that a parallel port is available. The phasing out of the parallel port in new machines doesn't mean you will not be able to use it off into the future, just that it may become more difficult to find a computer with one, or it may be necessary to buy and install a parallel port card in a computer that doesn't have one. Once you have found a suitable computer with a parallel port, you should be able to use it pretty much indefinitely as a controller for your CNC machine.

The USB Port

The USB (Universal Serial Bus) port will probably eventually replace the parallel port as a conduit for step and direction signals. USB-based designs may avoid some of the drawbacks of the parallel port, such as the limitation in the speed at which the parallel port may operate, and some of the wiring hassles. Unfortunately there is little to report because it does not appear that a good, affordable USB based option is yet (as of this writing) available for hobbyists.

The Controller (Controller Box)

The primary function of the controller is to precisely control the motors of a CNC machine according to the signals it receives from the PC which are generated by the controller software. Additional features may include the ability to control different devices that are part of the machine, and a means to stop the machine in a hurry if things go badly.

A basic controller box typically contains at least a few of the following components:

- **Motor drives,** to provide circuitry to control the motors on the machine. Motor drives that interface with the parallel port operate using 'step and direction' signals, which are described below. Motor drives for controlling servo motors also receive feedback from encoders.
- **A power supply,** for converting the AC voltage from the outlet to a specific DC voltage (or voltages) for use by components of the controller box, such as motor drives and relays. The power supply is an important enough aspect of a motor controller system (and many other electronic devices) that it will be covered in its own chapter.

- **Fusing** to keep electrical faults from turning into a major catastrophe.
- **Relays** to switch on and off different devices on the machine.
- **A breakout board** to allow wiring access to parallel port pins on the PC for input and output purposes (as described previously).
- **An E-stop** (emergency stop), which is an easily located and activated switch which will stop the machine quickly in critical moments.
- **Switches, connectors, wiring and other hardware**, which are used for their standard purposes.

Motor Drives

A motor drive is a relatively complex electronic circuit that is designed to control the movement of stepper and servo motors. The motor drive is an intermediary between the computer and the motor, and is responsible for translating the control signals sent by the computer running the controller software into appropriate current and voltage to precisely control motor movement. Motor drives such as the ones shown in Figure 22 operate via 'step and direction' signals.

Step and Direction

The phrase 'step and direction' refers to a system of control signals used with motor drives. Each motor drive must be wired to two output pins of the parallel port. One pin is used to send a signal to the motor drive to tell it when to move a step (the smallest possible increment), and another pin is used to send a signal to tell it which direction to turn.

How fast a motor spins is dependent on the frequency of the step signal sent to the motor drive. The output pins of a parallel port can produce an 'off' or 'on' (1 or 0, high or low voltage) signal, a.k.a a digital signal. Each digital pulse sent to the 'step' input of a motor drive tells the motor drive to move the motor one increment, or step. The faster the step signals are sent to the motor drive telling it to move a step, the faster the motor will turn (see Figure 21), up to a maximum.

The direction that the motor turns depends upon the signal sent to the 'direction' input of a motor drive. When the motor is to reverse direction, the signal changes from whatever state it is in, to the other. The state that represents clockwise or counter-clockwise rotation is arbitrary, however. It may be set up in the controller software.

Controller Box Examples

Figures 19 and 20 show a couple of different controller box examples. The one shown in Figure 19 is a hobbyist design meant to control servo motors and the one in Figure 20 is a commercial unit for stepper motors. Both use motor drive modules manufactured by the same company which look similar to each other, but the internal circuitry and the hookup is different.

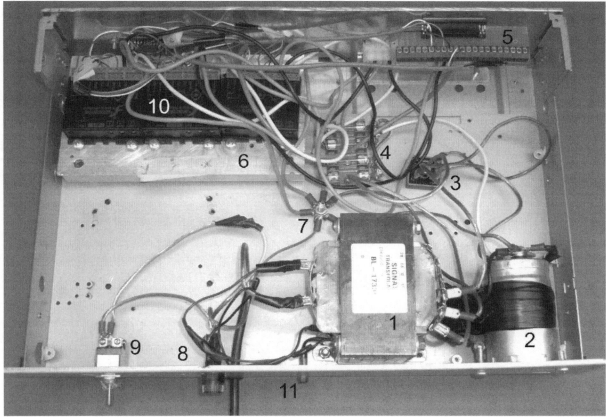

Figure 19: *A DIY controller box: Not the prettiest (or best conceived) layout in the world, but it does show the essentials. 1) transformer, 2) capacitor, 3) rectifier, 4) fuses (for the motor drives), 5) breakout board, 6) heat sink, 7) star ground, 8) fuse for incoming (mains) power, 9) power switch, 10) motor drives (one for each axis), and 11) Power indicator light. The motor controllers require a source of 5 volts to operate the low level (i.e. transistor) circuitry in the motor drives. To avoid an additional power supply, this box poaches 5 volts from the PC it is connected to through the USB port.*

A simplified wiring diagram is included in Appendix E. For a more polished and feature laden box, see Figure 20, below.

Figure 20: A *controller box (manufactured by Denver CNC). This is a four-axis controller box and contains four motor drive modules instead of three as in the previous example. Additional features beyond the basics include an E-stop (the button at the bottom of the picture), temperature sensors to protect the drives, and relays for turning external devices on and off. The rear of the box contains a fan and connectors.*

As mentioned elsewhere, stepper motors and servo motors operate on different principles, and the circuitry in each type is substantially different. However, if a motor drive says that it operates using step and direction signals, then it should be relatively compatible with any controller software that outputs step and direction signals, regardless of if it is a servo motor drive or a stepper motor drive.

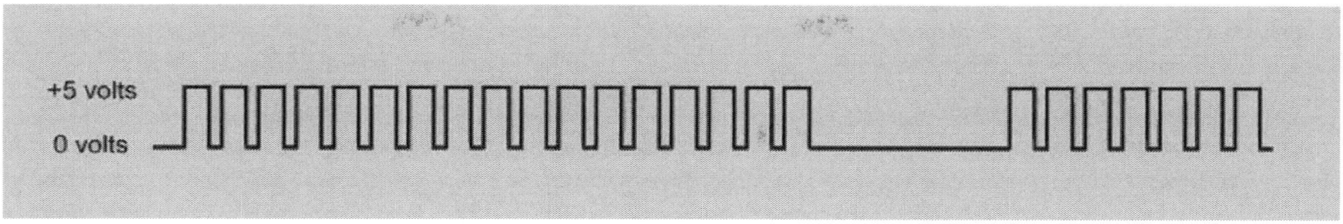

Figure 21: *A stream of digital pulses. Each pulse instructs the motor drive to move the motor a small increment. Repeated pulses produce continuous motor rotation, where a lack of pulses causes the motor to hold in position. The pulses are produced by a printer port in a range of approximately 25,000 to 65,000 times a second. The maximum value depends largely on the speed of the computer being used and how much software is running in the background.*

Factors Influencing the Cost and Performance of Motor Drives

Motor drives can vary greatly in price, and given that most CNC machine use three or more, cost is a big consideration. Factors influencing the price include:

- **The maximum current capacity of the motor drive:** drives with higher current capacities generally cost more.
- **The ability to microstep** may come at a premium.
- **Pulse multiplier circuitry** may also come at a premium.
- **The overall quality of the circuitry:** Not all driver circuitry is created equal. Some may have better designed and more sophisticated circuitry that is worth paying more for.
- **Kits** tend to be cheaper than fully manufactured drives, but you earn your discount with 'sweat equity'.

For smaller machines running motors with relatively small current demands, less expensive motor drives may work perfectly well. Obviously, larger machines running larger, more powerful motors may need high current motor drives, and these usually cost more. It is important to determine the current requirements for a given motor before purchasing your drives, however, because underpowered drives will not be able to get full performance out of motors with a high current rating. It would probably be wise to examine the choices of people who have converted or built a CNC machine like the one you want to build to see what drive and motor combinations have been used successfully.

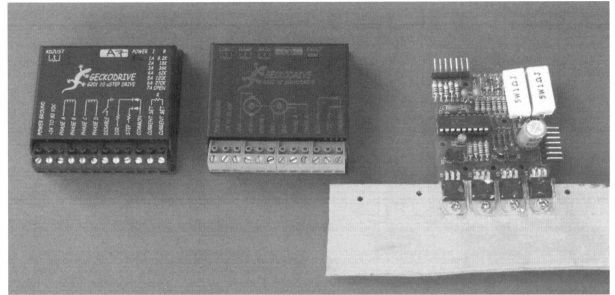

Figure 22: *Three motor drives: a bipolar stepper motor drive (a Geckodrive), a permanent magnet servo motor drive (also a Geckodrive), and a kit-built unipolar motor drive (Linistepper). All three use identical control signals (step and direction). The servo motor drive requires feedback (from an encoder) to operate, where the two stepper motor drives do not. There are a lot of options available these days, and more are being created, so shop around.*

The quality of the circuitry used can play a large factor in determining the overall performance of a system. Some drives are engineered better than others, and will produce greater torque and smoother operation driving the same set of motors. Also, some drives fail more often than others. Specs don't tell the whole story. Do your homework before purchasing a set of drives, since they are relatively expensive and a pivotal component in a CNC machine.

Microstepping and Pulse Multiplier Circuitry

Microstepping circuitry and pulse multiplier circuitry are enhancements used in stepper and servo motor drives respectively. They can increase the flexibility in setting up a CNC motor system, and provide substantial performance benefits as well. As described earlier, they come at a price, but may be useful or essential in some situations. Their necessity is dictated by the particular application for which they are being used.

The term *microstepping* refers to the use of additional circuitry to allow the motor drive to turn a fraction of a step. For example, if a stepper motor which requires 200 steps to make a whole revolution is being driven by a stepper motor that is operating in 10 microstep mode, then there will be 2000 steps to turn the motor a full revolution.

The ability to microstep is useful because it improves the resolution of a CNC machine; it allows for more precise movements. Microstepping can also improve the smoothness with which stepper motors operate. Aside from added cost, it isn't a bad thing to have microstepping available in a stepper motor drive. With most drives, you can choose to use it or not if you want. Some drives allow selection of several different microstepping modes, from no microstepping to half stepping (two half steps per full step) all the way up to whatever the maximum number of microsteps is available, such as ten or sixteen microsteps per step. Microstepping is specific to stepper drives, and is not relevant to servo drives.

Pulse multiplier circuitry is designed to create multiple step pulses from a single input pulse. This can be used in servo motor systems to achieve greater speed, and also allows the use of encoders with higher than normal resolution. These issues are detailed in the following section.

Encoders

If you are considering using servo motors, then some knowledge of how encoders work is in order. An encoder is a feedback device which provides information about linear or rotational motion. Servo motor drives require the use of feedback to provide information about how much a motor has turned. Encoders are not part of the controller box per se (they are usually attached directly to the servo motors) but they are an integral part of the control electronics in a servo system.

How an Encoder Operates

There are a number of different types of encoders. The encoders of greatest interest to a CNC hobbyist are called rotary encoders, and provide feedback in the form of electrical pulses, which are generated as the shaft of a motor is turned. Encoders commonly use a finely slotted disk and optoelectronics, i.e. an LED light source and a photo detector inside the encoder. As the disk in the encoder rotates, the light passes through the slots and triggers the photo detector, producing a stream of electronic pulses as it turns. These digital pulses provide the motor drive with an indication of how much a shaft has turned. The motor drive circuitry produces movement by sending current to the servo motor, and then keeps track of the number of pulses it received back from the encoder.

Important terminology relating to the use of encoders includes the following:

- **Resolution** characterizes how small a movement an encoder is capable of detecting.
- **Quadrature:** This term refers to a mode of encoder operation that is designed to enhance the resolution of an encoder and also indicate the direction of rotation as well.
- **Style of encoder:** Encoders are used in a wide variety of applications, and different styles of encoder have been created to suit.

Resolution

Resolution refers to how small a movement can be detected by an encoder, and is a fundamental characteristic of the encoder. For the type of encoder used in most DIY CNC machines, resolution refers to how small a fraction of a revolution may be detected (and hence how small a movement a motor can make). This directly influences not only how small a movement may be made by a CNC machine, but the maximum speed of linear travel achievable as well. The relationship between speed and resolution is discussed in Chapter 12.

Motor drives may have restrictions on the resolution of the encoders that they will work with. Different motor drives have different minimum and maximum resolution requirements, and drive manufacturers should provide this information in a technical document or manual. Before purchasing encoders (either ones that come with a servo motor or separately) verify that they meet the resolution requirements of the motor drives you will be using. Also keep in mind, that even if a motor driver will work with a motor drive, the resolution can have a big impact on the maximum speed of travel. This is covered in Chapter 12.

Quadrature

Quadrature refers to the use of a pair of output devices in an encoder to produce a pair of out of phase pulse streams. These two pulse streams are compared in motor drives designed to work with quadrature encoders to provide two benefits; enhanced resolution and the direction of rotation.

For instance, in a 256 CPR (counts per revolution) encoder there are two optical detectors that will each produce 256 electrical pulses during a single rotation based on the 256 slots in each disk. Circuitry in the motor drive compares the two pulse streams, providing both the direction of rotation of the motor and a fourfold increase in resolution. This results in 1024 increments per single rotation of the motor shaft. This process is a little like how the brain compares the images it receives from each eye to gain improved vision and depth perception.

Figure 23: *At left is a servo motor with an encoder attached at the back, alongside a panel mount encoder (right). Rotation of the servo motor shaft causes the encoder to produce a stream of digital pulses, as does rotating the shaft of the panel mount encoder. Both of these devices produce similar output, and have a similar hookup, but have different packaging.*

It is not absolutely necessary to understand exactly how the process works, but remember that a quadrature encoder has four times the resolution of the listed CPR (counts per revolution) value when matched with an appropriate motor drive.

An encoder with quadrature outputs typically employs two wires for the pair of channels, another wire as the ground (zero volts), and another wire as a 5V source for the encoder electronics, giving a total of four wires. The two out of phase channels are commonly marked as A and B or something similar. Encoder and motor drive manufacturers will supply technical references and manuals detailing the wiring and operation of their products.

Encoder Styles

There are a variety of different encoder styles and types to fit the variety of uses for this device. Linear encoders are designed to measure straight line movement directly, such as with a digital caliper. There are also rotary encoders that are designed to measure absolute position (angles) in a complete circle (a 360° arc). Some differences between encoders are strictly application specific. A through-hole encoder is designed to have a shaft of a motor passed through it to sense rotation of the shaft. Panel mount encoders are designed to be mounted to the front face of a piece of equipment to act as an input device. Both of these devices may use nearly identical electronic and mechanical devices to create nearly identical output (see Figure 23), for the identical purpose of precisely indicating how far a shaft has been turned.

9

Electronics: Power Supplies

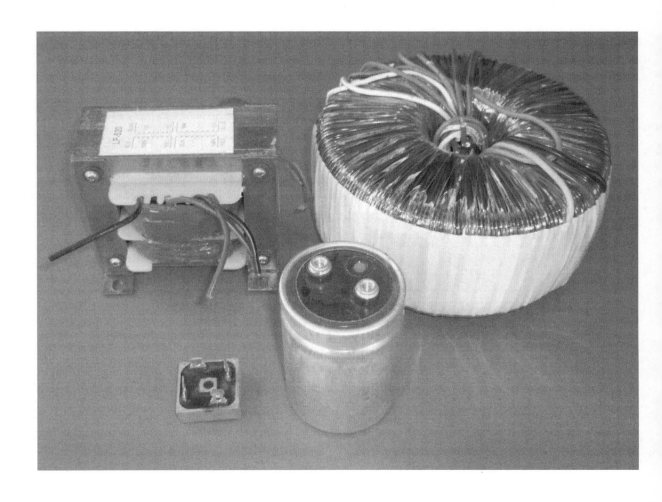

Yet another warning / disclaimer / stating of the obvious: Electricity, especially at the voltages delivered at a wall outlet, can be very dangerous if not approached with caution. If you are not familiar with the nature of electrical devices such as transformers, capacitors, and operation at mains voltages (the voltage supplied at the wall outlet), then make sure you get assistance from someone who has appropriate experience. Your heart is, in fact, an electrical device as well as a muscular one, and exposure to high voltages may stop it permanently. A very basic introduction to electricity and electronics is included at the end of the book.

What is a Power Supply?

A power supply is a device that takes the electricity that is pumped through the electrical grid to our houses, offices, and factories, and modifies it until it is in a form that is useful for a particular purpose. For instance, in a personal computer the power supply receives 110 Volts AC, (220 Volts in Europe) and converts it into a variety of voltages (12 volt DC, 5 volt DC, etc) that are useful for powering the circuits and hardware of the computer.

There are several different types of power supply. One of the most common is a *switching* (or switched) *power supply*. This type of power supply is fairly complex, and is usually purchased pre-built. Another type is referred to a *linear power supply*, and is relatively simple compared to a regulated power supply. This type is simple enough to be (and often is) built by do-it-yourselfers. It is useful to understand the parts of a basic linear power supply, regardless of whether you will build one yourself. The most important components are:

- A **transformer**, to change the incoming alternating current to a different voltage, more suitable for the application at hand.
- A **rectifier**, to take the alternating current from the transformer and turn it into direct current (DC).
- A **capacitor**, to smooth out the fluctuations in the rough DC output from the rectifier to acceptable levels.
- **Fuses** provide protection against situations where the power supply or component that it is powering draws more current than it should under normal operation. This is an extremely important component, and can prevent damage to electronic components, and more importantly prevent fires and electric shock.

Figure 24 (on page 83) *shows some of the parts found in a linear power supply. The back row is composed of two different styles of transformer; a laminated core transformer and a toroidal transformer. In the front row are a rectifier and a large electrolytic capacitor.*

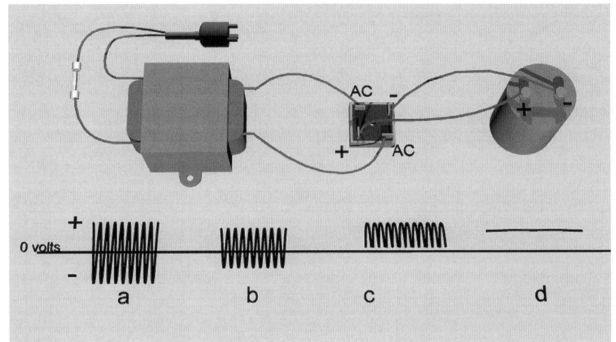

Figure 25: *The power supply receives current at full voltage from the power outlet to its primary leads (a), and the transformer changes it to a different voltage (b) at its secondary leads. The transformed voltage is then fed to a rectifier which produces direct current (i.e. only positive or negative) voltage (c) which is very choppy. This signal is fed to a capacitor which smoothes out the bumps producing a relatively constant voltage. Please note that the capacitor in this case is usually a high capacitance, electrolytic capacitor that has a specific polarity. If it is not hooked up properly (i.e. the positive output from the rectifier goes to the positive lead on the capacitor, and the negative output from the rectifier goes to the negative) it may explode when the power is first turned on. For simplicity this figure excludes a ground wire that would normally be connected to the case and provide protection against electrical faults that might occur.*

In general, the more current and voltage that a power supply is required to supply, the larger and more expensive the components to create it will be. Transformers require substantial amounts of wire for their operation, and transformers that handle large amounts of current require large amounts of thick wire, increasing their weight and expense. High capacity power supplies also tend to incorporate large value capacitors (known as filter capacitors) to adequately deal with the large amount of current.

Types of Transformers

It is worth mentioning the two common types of transformers available for use in power supplies, standard (laminated core) transformers, and toroidal transformers. Both transformer types perform the same function, but toroidal transformers provide a couple of benefits relative to standard transformers.

Toroidal transformers are a bit more efficient, and produce more power for a given transformer weight; they also produce less electromagnetic noise than laminated core transformers. Toroidal transformers are favored in some applications where low noise is desired, such as high quality stereo equipment and power supplies for medical equipment.

Both types of transformer may be used effectively for power supplies designed to turn CNC motors, however. It's fine to use either type. It is possible to reduce the electromagnetic noise produced by a transformer through the use of metal shielding if necessary.

Transformer Terminology

Transformers come with two sets of wire leads, known as *primary windings (a.k.a. primaries)* and *secondary windings (a.k.a. secondaries)*. Primaries are the leads that are hooked up to the alternating current from the power grid; the wires that will be connected to the power cord that get plugged into the wall. Secondaries are the output wires; they produce the transformed voltage that is to be fed to the rectifier and transformed into direct current.

Transformers may have more than one set of primary wires to allow flexibility in wiring of the transformer. For instance, extra primaries may allow a transformer to be wired for either 220 Volts (as in Europe) as an input, or for 110 Volts (as in the United States). Transformers may have as few as one pair of wires for the secondary voltages or they may have many additional wires providing a variety of different voltages at different current capacities.

The term *center-tapped* indicates that a secondary has an additional lead connected to its center (midway between its ends). This allows the creation of dual voltage power supplies (i.e. with positive and negative voltages of the same absolute value) which is useful for certain applications. If a transformer has dual secondary windings, this means that there are two independent coils which will produce the same secondary voltage. A transformer with multiple secondary windings should have the voltage and current capacity of each set listed individually.

Other important terms to know when evaluating transformers are *step-up* and *step-down*. Step-up transformers take the input voltage and raise it from the input voltage. Step-down transformers take the input voltage and drop it to a lower voltage. In the types of applications discussed in this book, step-down transformers are used.

There are also transformers that take the incoming voltage and turn it into the same voltage. This may seem pointless, but this is a way to clean up the irregularities of the incoming voltage supply for situations where high stability is necessary. These are known as *isolation* transformers.

How much of a Power Supply is Necessary?

A common question for someone who is building up a CNC controller box is that of 'How much?' How much current should it be able to supply? What voltage should it be? These questions depend largely upon the requirements of the motors and the motor drives. Motors and motor drives have different maximum and minimum voltage and current requirements. Too much voltage can result in damage to a motor drive or motor. Too little voltage or current can result in poor performance or no performance at all.

Determining Current Requirements

A first approximation of current requirement would be to determine what the maximum possible current draw is for each motor during operation, and sum them up to get a value. This is actually an upper limit, and although it is a useful number to know, it is probably substantially more than is necessary in practice.

During operation the machine is typically only driving a couple of motors at a time, and of those that are being driven it is unlikely that they are all being driven near their limits. It is highly unlikely that all motors will draw near their maximum current simultaneously. A CNC controller box will also draw a small amount of power for operation of supporting components, but this amount is fairly small and stable when compared to what the motors draw during operation.

That being said, it is still a good idea to have a certain amount of capacity beyond the average expected current draw. One manufacturer of stepper controller boxes I consulted felt that 60% of the maximum possible current draw from all devices was a relatively safe figure. There do not appear to be any absolute guidelines, however, and having a little extra headroom won't hurt.

What Voltage Should my Power Supply Be?

The voltage of the power supply for a CNC controller box is determined by a couple of constraints; primarily the voltage ratings of the motor and the motor drive. For instance, if the motor drives you are using can accept an input voltage in the range of 25 Volts to 75 Volts, then the power supply must provide a voltage in this range. If the motors you are using with these drivers have an upper limit for operation of 45 Volts, then it would be pointless (and damaging) to use a power supply above this value. In this case, a voltage between 25 and 45 would work, with higher voltages providing a greater top speed for the motors.

Essentially you should create a power supply that falls within the lower limit specified by the motor drive and the upper voltage limit for the motor itself, so long as it is less than the maximum allowable voltage for the motor drive. In the case that the motor's maximum operating voltage exceeds the motor drive's maximum voltage, the upper limit is the motor drive's upper limit.

Determining a Motor's Maximum Operating Voltage

With a stepper motor, the motor may be run at its specified current value at a value anywhere from approximately 5 to 15 times the rated voltage. This should only be done with a stepper motor drive that has current limiting capacity or through the use of ballast resistors to limit the current in cases where the motor drive lacks built-in current limiting. Current limiting is a requirement because without it, stepper motors will draw excessive current at higher voltages, and will be damaged otherwise.

For instance, with a stepper motor specified at 3 volts and 3.5 amps, the voltage the stepper motor may be run at is anywhere from approximately 15 to 45 volts, with the current set at a value somewhere below the rated value of 3.5 amps. There is some variation in what is considered an acceptable upper limit voltage. Pushing the voltage up higher results in greater heat production, however, which can damage motors if not dissipated rapidly enough.

With servo motors, the upper voltage limit should be taken from the manufacturer's specifications if they are available. If, for instance, the maximum voltage is 40 volts, then stay below this value. The minimum voltage will probably be determined by the minimum voltage requirements of the motor drive used with the servo motor.

Multiple Voltages

Motor drives and devices like relays usually require that additional supply voltages are available. Some motor drives, for instance, require an additional logic level (approximately 5V DC) voltage supply. The controller box in Figure 20 for instance, has a 40 Volt unregulated supply, and a 12 and 5 Volt regulated supply. The 40 Volt supply is for the motors, and the 12 Volt and 5 Volt supplies are for triggering relays and to supply to logic level circuitry in the drive, respectively.

To get additional voltages, separate power supplies may be used (as in the controller box in Figure 26), or additional circuitry (such as a voltage regulator circuit) may be used to create a different voltage from one already available. Another possibility is to 'borrow' from another device as done with the controller box shown in Figure 19. It has an unregulated supply of about 30 Volts, and poaches 5 Volts from the computer by way of the USB port. This was done to avoid creating additional circuitry necessary for producing the regulated +5V used by the motor drive circuitry.

Unregulated and Regulated Power Supplies

Regulated power supplies incorporate circuitry to maintain very tight control over the voltage produced by a power supply. Unregulated power supplies lack this additional circuitry, and rely on the quality and size of their components (and the quality of the electricity they get from the power company) to maintain a reasonably stable voltage. They are used in situations where some fluctuation in voltage is not a problem and where large, instantaneous current demands are possible. A CNC controller box may use both types; unregulated power to supply the motors it controls, and a regulated supply for the motor drive circuits. This may be done by regulating a portion of the unregulated supply, or by installing a separate regulated power supply.

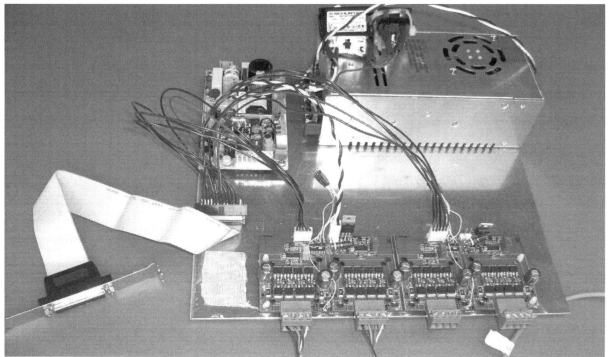

Figure 26: *A prototype of a four axis motor controller box. This box uses a switching power supply (the large box at the back right) purchased from a surplus electronics supplier to provide current to the motor drives (foreground). Another pre-built supply (back left) was purchased surplus to provide other voltages required by other devices, such as a 12V and 5V used by relays and the motor drive circuitry. The motor drives (for four axes) are shown in the fore.*

Switching Power Supplies

A switching power supply uses some complex circuitry to substantially improve the efficiency of a power supply when compared to a linear power supply. The advantages of switching power supplies are light weight and in general, lower manufacturing cost for equivalent power. They achieve this by directly rectifying the incoming wall voltage.

Drawbacks of a switching power supply relative to the linear power supply include increased noise (i.e. electrical interference), and greater complexity. Both these issues are not a great problem, however. The complexity issue is solved by not actually building it; if you see one that suits your needs, just buy it. There are a lot of different ones available, and they may often become available as surplus. Potential noise issues may be corrected by shielding the power supply.

Another more significant issue, however, is that switching power supplies do not handle instantaneous power demands produced by the motors as well as linear power supplies, and in general, aren't particularly suited to meet the current demands presented by motor drives (and the motors they are driving). They should be somewhat overrated (have a higher power rating) than a linear power supply when used as a supply for driving motors.

It is quite possible to use switching power supplies in a CNC project to supply the motor drivers. However, linear power supplies tend to be simpler and, in general, work reliably. Their simplicity also means that they may be built up from surplus parts to suit the needs of a given project. In general, it may be better for novice builders to stick to a linear power supply for supplying current to drive their motors, because using a switching power supply can be a bit tricky in some situations.

Fuses and Circuit Breakers

Fuses are relatively simple, but extremely important devices. A fuse is usually just a strip of metal or wire encased in a housing that is designed to burn up and create an open circuit when a specified current level is exceeded. This is why they are specified in Amps. There are a variety of different package-types (shapes and sizes) for different types of applications, so there are many different choices available.

For someone creating their own power supply from the parts describe previously (transformer, rectifier, capacitors, etc) a fuse or some form of circuit breaker must be used to protect against any electrical faults that might occur. For instance, a short circuit within a controller box could cause the production of heat or sparks resulting in a fire unless a fuse cuts off the flow of current. Most power supply diagrams specify some sort of fuses in the design. Don't ignore them.

Fuses may also be used to protect various components within a device. If there is some component that you want to make sure doesn't burn up from drawing too much current, then placing a fuse in line with its power supply can provide a way to do this.

Many motor drives have a current limiting feature which allows the maximum current draw to be set on the driver. This restricts the amount of current that the motor may receive from the motor drive, protecting the motor from damaging levels of current. Even in this case, however, a fuse may be used to provide backup protection if the motor drive fails in some other way during operation and current limiting is lost.

Slow Blow Fuses

A slow blow fuse differs from a regular fuse in that it can absorb very short bursts of current at or beyond its rated value. Regular fuses will blow (break a circuit) as soon as their rated current limit is exceeded, but slow blow fuses will not. Slow blow fuses are useful in situations when a certain current limit is desired, but a small, short term excursion above that limit may occur during normal operation and is acceptable.

An example would be where one is used as a way to deal with current inrush when a power supply is turned on. The initial high current flow diminishes rapidly as the capacitors in the power supply are charged. The inrush current is higher than desired for normal operation of the device, but is acceptable for the short (and predictable) brief period after the supply is turned on. If a fast blow fuse were used, it would be necessary to use a higher value than the slow blow fuse so that it could withstand the inrush, but this value may be higher than desired for continuous operation.

Circuit Breakers

Circuit breakers perform the same function as fuses, (i.e. they will break a circuit once a current limit is exceeded) but are resettable and may be used repeatedly without replacement. Like fuses, they are also an incredibly important safety device, so if you intend to use them, you should understand a couple of things. Fuses may have considerably different ratings when used with AC or DC. Circuit breakers will commonly have high AC voltage ratings (for instance they may be usable up to a maximum of 250 Volts AC), but this does not mean that they will have a comparable DC rating (it may not be rated for use with DC at all). Make sure that you know what the specific voltage rating of a circuit breaker is prior to using it in your power supply or other device. Also, circuit breakers with low current ratings (a few amps or so) tend to be fairly expensive, so this may restrict their usage. For some situations where you are prone to blow the circuit on a somewhat regular basis, it might be worth the investment. In general, fuses are relatively cheap and easy to implement, so they would probably be the first choice in most cases.

10

CNC Conversions

Converting an existing machine is one of the quickest ways to get started in CNC. I have chosen to examine the conversion of a couple of different bench-top mills from manual to CNC control, since many of the people reading this book are interesting in doing a mill or lathe conversion as a first (and perhaps only) CNC project.

Although small scale machines are covered here, much of this information should still be useful for converting larger machines because many of the same principles apply. The conversion or retrofitting of mid-sized and larger mills to CNC operation is not uncommon for a hobbyist. In general, it is a matter of scaling up. The motors, electronics, and hardware tend to get larger and/or more powerful.

The basic idea when performing a CNC conversion is to take an existing machine and successfully attach motors to it to drive the screws on it. In some cases, it may be necessary or desirable to replace the screws as well to get acceptable (or exceptional) performance.

A show-and-tell approach will be taken in this chapter, using case studies of some CNC mill conversions, describing how various design decisions were made. The first mill described is a Proxxon mini mill (the MF-70).

A Proxxon Conversion

The Proxxon is a small machine (approximately 5" X 9" X 13.5") that is intended for light-duty applications such as model making or cutting wax for jewelry making. The table is approximately 8" long by 3" wide with a travel of approximately 5 inches by 2 inches. It incorporates a variable speed spindle on the Z-axis, with a travel of about 3 inches. The machine was not designed with CNC conversion in mind, so substantial creativity was needed to properly mount the motors to the machine. The machine came with hand-wheels and plastic collars for manual operation of the axes, which were removed to reveal the ends of the lead screws that are used to position the mill table.

Selection of Motors

For this mill conversion, a choice of motor was made based on power and weight considerations. Given the relatively light weight mill table on this machine, and the fact that the motor hangs off the edge of the long axis (the X-axis) a significant

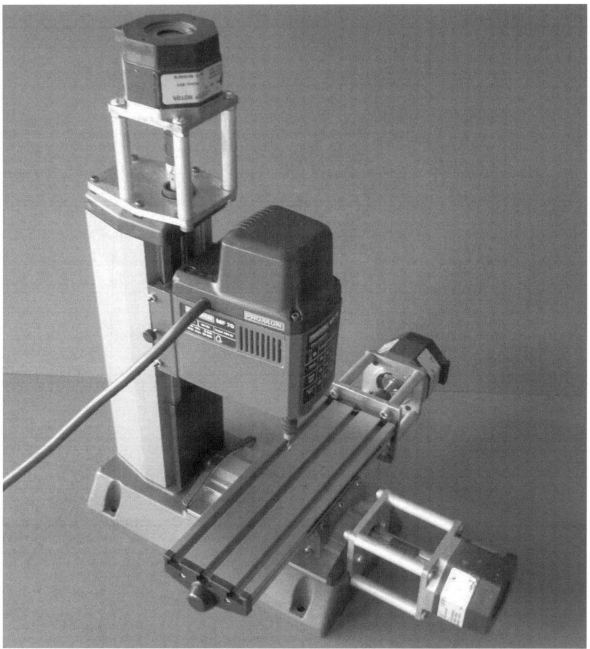

Figure 27: *A profile of the Proxxon micro-mill conversion. The original machine wasn't designed with CNC conversion in mind, so several parts had to be fabricated to properly attach motors to the machine.*

distance, the lightest motor that had sufficient torque for the job was needed. Heavier motors would make the small mill table bind up, wear prematurely, and would probably cause the table to loosen up frequently during use. The motors shown in Figures 27-29 are relatively short 'single stack' motors and were the smallest motor size that could drive the axis adequately without stalling.

The motor used to drive the Z-axis could potentially be a larger, higher torque motor if necessary because the orientation of the motor does not create the same kind of problems seen with the motors that hang off the ends of the mill table. In this case the single stack motor was sufficient to move the spindle up and down without problems.

It would be possible to use small servo motors (such as the one pictured in Figure 23) to drive this mill, although it would be necessary to gear them down in some way to accommodate the torque and speed characteristics of these motors. The mounts might be a bit more difficult to make as the motors don't have the convenient mounting holes built into the stepper motor bodies. Either servo or stepper motor would probably work, though.

Motor Mounts
Three mounting plates were cut to allow the motors to be securely attached to the three different axes. The main features in these plates are a set of holes to match the holes in the motor frames, additional holes for the screws that secure the plates to each of the three axes, and some additional machining to allow the plates to mount flush to the ends of the table and top of the spindle. For each motor mount, aluminum tubes were used as spacers along with bolts to mount the motors enough of a distance so that the shaft of the lead screw and shaft of the motor did not collide. An adapter was created for the end of the motor shaft which had a diameter that was the same as the motor shaft, and the motor and the screw were coupled with an adapter as shown in Figure 28.

It is possible in a configuration such as this for the motor to twist somewhat at the ends of the aluminum tubes during operation. This conversion uses relatively large diameter spacers, is strongly bolted, and uses motors that aren't terribly powerful, so this construction should be more than rigid enough to resist this potential problem. With a heavier duty machine which will generate greater forces than this one, this design might be questionable.

Coupler and Shaft Adapter

The motor has to be connected to the shaft of the lead screws for each axis in some fashion, and in this case a coupler was purchased from one of the suppliers listed in Appendix B. This particular coupler is composed of three parts, two metal hubs and an X-shaped rubber spider in between the hubs to allow for slight misalignment between the motor and the shaft being driven. Set-screws are used to hold the coupler to both the motor shaft and the lead screw/adapter.

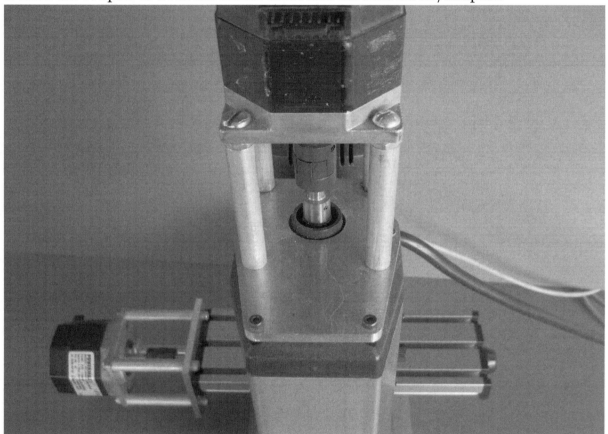

Figure 28: *A close-up of the Z-axis mounting plate, coupler, and lead screw adapter (from behind the machine); the plate was cut and drilled to mount flush to the top of the Z-axis, and holes were drilled to use already existing bolt holes on the cap of the Z-axis. Proper alignment of the motor shaft and the lead screw is very important.*

An additional adapter was created out of brass using a lathe to create a shaft end that was an appropriate diameter (.25 inches) for use with the coupler. The creation of motor mounts and associated parts like the shaft adapter requires considerable care (they must be highly precise and concentric) because proper alignment between the motor shaft and motor is essential for proper operation.

Figure 29: *The X-axis motor mount, stepper motor, coupler and lead-screw adapter. A small motor was desired because a heavy one could cause the relatively small mill table to bind up, and might cause it to loosen up rapidly during use.*

Final Comments

This is a nicely executed conversion (sadly, I can't take credit for it). The difficulty in converting the Proxxon came in how to mount sufficiently powerful motors to the relatively small and light-weight mill table. Additionally, proper alignment and connection of the motor and motor shaft required that relatively precise mounting plates and an adapter be made. This is no small task if you do not have access to the appropriate machinery, or know someone who does. A solution used on other conversions of this same machine was to use a piece of plastic tubing (available at home improvement stores) to act as a coupler between the motor and shaft. The flexibility of the plastic tubing allowed it to fit over both the motor shaft and the lead screw even though they had slightly different diameters, and it also compensated for misalignment between these two shafts. This type of coupler, however, is only suited for very low power, light duty situations.

The Devil is in the Details

Some of the issues in this conversion, such as the need for a relatively light motor on the X-axis, are specific to this machine. Other issues, such as the significant effort in properly aligning and connecting the motor shaft and lead screw are problems that are general to any home built machine or conversion of an existing manual machine. Finding an appropriate coupler, creating rigid, properly aligned motor mounts (in this case, the combination of metal plates and tubular standoffs) are also pretty standard hurdles to overcome. Although not a problem with this conversion, some mills have a Z-axis that is quite heavy, and may require a more powerful motor than used on the other axes. Also, the Z-axis on some machines may require the addition of weight to balance the load on the Z-axis lead screw.

A Sherline Conversion

The Proxxon conversion was relatively challenging, requiring substantial time and effort in to the design and creation of needed parts. A much easier road is shown in the following walkthrough of a Sherline mill conversion using a kit from the manufacturer.

For those unfamiliar, Sherline produces a collection of popular bench-top scale mills and lathes. This is a popular CNC conversion and was included for those who are interested in this particular brand of machine and conversion kit. Readers who aren't specifically interested in this may skip this section if desired, although it wouldn't hurt to scan the pictures to get a feel for how the conversion is done. The basic idea is usually the same with different conversion kits: remove some original hardware, attach motor mounts, and then mount the motors and couple them to the drive screws. In some situations, a drive screw or screws will need to be replaced.

This conversion is done on a Sherline 'eight-way' mill, and uses a conversion kit manufactured by Sherline. The eight-way mill is a fancier version of their standard mini mill, allowing additional adjustability in set up of the Z-axis. The conversion to CNC, however, is largely the same as with their standard mill.

Purchasing a conversion kit, if one is available for the machine you are interested in, makes the conversion of the mill to CNC operation considerably easier than designing and building your own. Many of the important details that go into engineering a decent conversion have been ironed out, and the manufacturing details are taken care of by a company (or individual) that most likely has good equipment available and practice in producing the parts. Many conversions can be done in a few hours with a kit if things go reasonably well.

The tools required for the conversion described here are easily available; hex keys, a drill, a clamp, a tap and tap holder, and a few others. If you don't have everything you need, it shouldn't be too hard to find the rest from a friend or neighbor. This conversion kit comes with three motor mounts (one for each axis), couplers to connect the motors to the lead screws, and a replacement lead screw for the Z-axis. These motor mounts are designed for use with NEMA 23 stepper motors (see Figure 4 for an example), which is a typical size for use with bench-top mills.

Figure 30-a

The Sherline mill being uncrated. The conversion kit is in the foreground, still in bubble wrap.

Figure 30-b

The first step after unpacking everything is to remove the hand wheels from the ends of the mill table using appropriately sized hex keys.

Figure 30-c

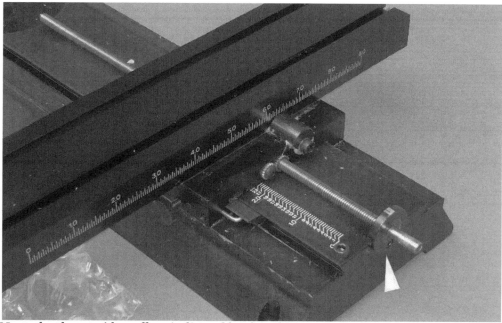

Next, the thrusts (the collars indicated by the white arrow in the picture) must be removed as well, using another hex key.

Figure 30-d

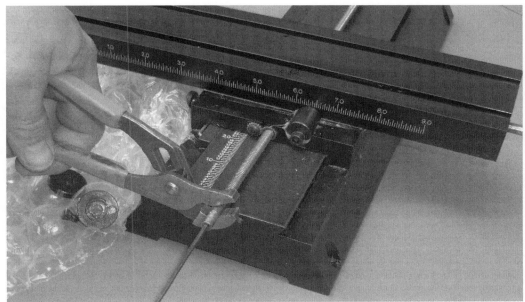

Another screw must be removed from the end of the lead screw. Pliers are used to hold on to a spacer connected to the end of the lead screw and a hex key is used to break it free and unscrew it from the end of the lead screw.

Figure 30-e

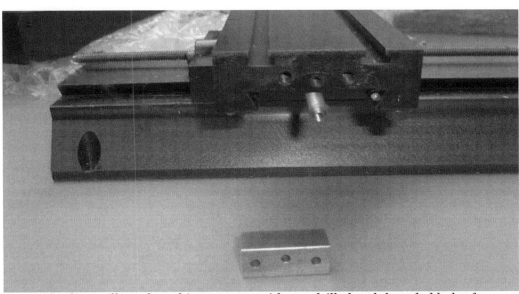

Newer Sherline mills such as this one come with pre-drilled and threaded holes for attaching the motor mounts. Sherline includes a metal block template (foreground) with properly spaced and sized holes for drilling older mills if needed. Refer to figures 31-n and 31-o for an indication of how to drill and tap the holes.

Figure 30-f

The motor mount is loosely bolted to the end of the mill table in preparation for the next step. The coupler (shown in inset) is mounted to the lead screw with an appropriate screw.

Figure 30-g

The lead screw is turned using the hex key until the table is moved to the limit of its travel. Doing this helps properly align the motor mount in relation to the table. The bolts holding the motor mount to the mill table are then tightened to secure it to the end of the mill table. This process is done for both the X and Y axes.

Figure 30-h

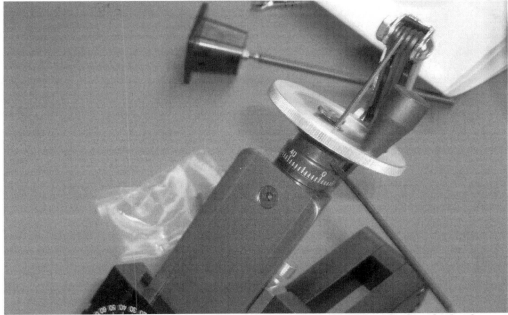

With the Z-axis, the first step is similar to that for the mill table. The hand wheel must be removed. In this case, two hex keys are used at the same time.

Figure 30-i

After the hand wheel is removed, the leadscrew thrust (collar) must be removed as well.

Figure 30-j

After the lead screw thrust is removed, the screw holding the saddle nut is loosened to allow removal of the lead screw. This is the only bolt in the previous picture and the top bolt in this picture. Above the saddle nut is the locking lever (the toe shape sticking out from the column bed), which will not be used in the CNC conversion.

Figure 30-k

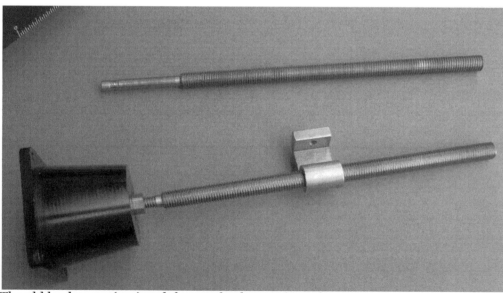

The old lead screw (top) and the new lead screw/ coupler / motor mount assembly included with the kit. The bronze saddle nut has been transferred to the new lead screw, but the locking lever will not be.

Figure 30-l

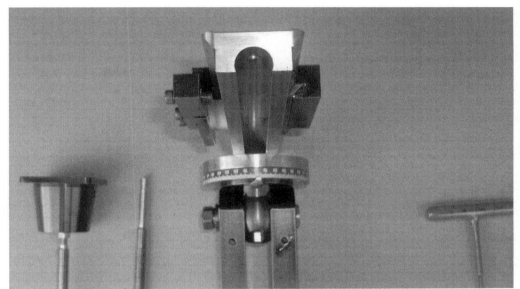

A top down view of the Z-axis column bed. The metal template for drilling the holes for the motor mount will mate with this semicircular cavity running along the length of the bed.

Figure 30-m

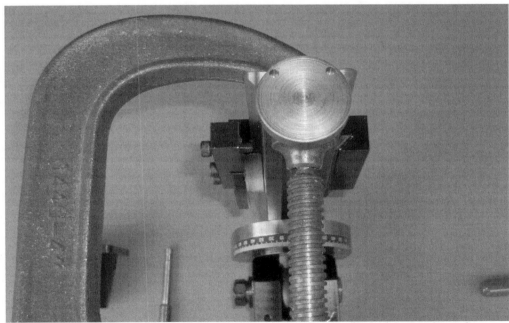

A C-clamp is used to hold the metal template in position. The holes are aligned as shown in the picture. Care should be taken when placing the clamp and screwing it down so as not to damage the metal on the column bed.

Figure 30-n

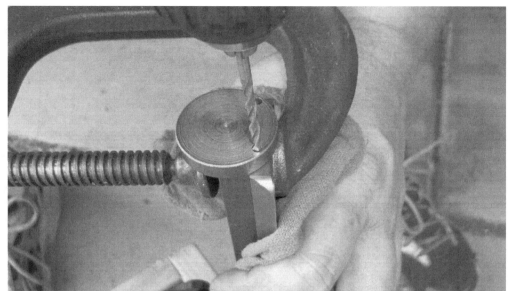

Holes are drilled with care taken to drill to the appropriate depth, also taking care to keep the drill bit vertical. Chips may need to be blown out periodically as the hole is drilled.

Figure 30-o

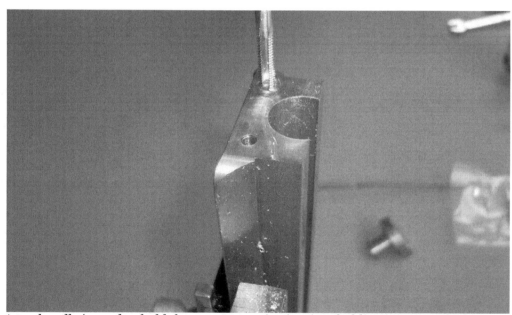

A tap handle is used to hold the tap. Improvisation (e.g. holding it in some locking pliers) may lead to bad results. The use of a lubricant such as WD-40 in the hole and on the tap can ease the tapping process. Chips are blown out of the hole periodically.

Figure 30-p

The motor mount / coupler / leadscrew assembly is bolted in position, and the saddle nut is bolted to the Z-axis sliding table.

Figure 30-q

When mounting motor, the flat portion of the motor shaft should be placed facing the set screw in the coupler. The set screw is tightened against this flat side to prevent slippage during operation. If the motor axis has no flat spot, make one using a file, or drill an indentation into the shaft to capture the set screw. A set screw pushed up against the round part of the shaft may slip during operation.

The rest of the mill assembly follows the instruction included with the mill. After all of this is finished, motors may be wired to the motor controller box, for some initial testing. The conversion kit makes the job of converting this mill very easy, not to mention that the parts are of very high quality.

CNC Lathe Conversions

CNC lathe conversions, like mill conversions, are accomplished by securely affixing motors to the machine and coupling them to the screws that are used for positioning. A detailed examination of a lathe conversion will not be conducted because the issues are largely the same as those seen with mill conversion, such as how to rigidly affix motors to the machine, and how to properly couple the motors to the screws. However, a couple of photos of a Sherline lathe conversion will be presented.

A Sherline Lathe Conversion

Below is a photo of the Sherline lathe CNC conversion partway to completion.

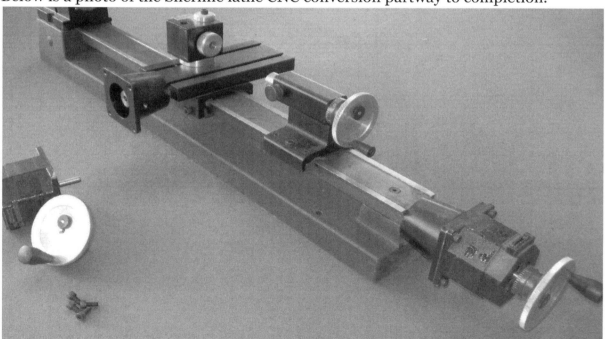

Figure 31: *A Sherline lathe conversion under construction. One motor has been mounted.*

This is the long version of their bench top lathe. As you can see, the conversion materials (mounts, couplers, etc.) are very similar to those used for the desktop mill.

Figure 32: *A side view of the same machine showing the motor mount for the cross slide table and its coupler. This conversion uses motors with a dual shaft (the motor shaft extends out both the front and back of the motor as shown in the un-mounted motor). This allows the original handles to be remounted at the back of the motor and used for quick manual positioning if desired.*

What Machines are Suitable for Conversion?

Note that some machines may not be suitable for conversion or are much more difficult to convert than others due to sloppy mechanisms, wear, and difficulty in attaching motors. Some of the main things to examine are the amount of slop (play) in the screws and nuts on the machine, how difficult it will be to make motor mounts, how the motors will be coupled to the machine's lead screws, and how difficult it is to turn the screws. These things are true for both mills and lathes.

Good candidates for conversion will have handles on the lead screws that are easy to remove, and leave a length of shaft available to allow a coupler or pulley to be mounted. Additionally, screws that are hard to turn will require more powerful motors, and probably electronics with greater capability to drive these motors as well. A screw with a lot of slop may require that the screw and nut be replaced. Although this adds cost and complexity to the conversion, it may present an opportunity to improve the performance of the machine by using better quality screws and nuts.

Synopsis

Of importance when converting a manual machine to CNC operation is which motors to use and how to attach them properly to the machine, regardless of the type of machine being converted. The motors need to be sufficiently powerful to move what they need to move and cut what they need to cut without stalling. They also must be mounted securely so they do not move or twist, to provide the most precise and accurate operation possible. Couplers must be located or created so that the motor shafts may be connected to and drive the screws on the machine. This sounds like a minor issue, but it can be a real sticking point in a home brewed CNC machine or conversion, because tight tolerances are required.

If you are looking to convert a relatively popular mill or lathe, then a conversion kit may be available, or some portion of a conversion may be available. Kits vary in completeness, ranging from something as simple as a motor mount, to a significant overhaul that includes replacement screws, motor mounts, motors, and so forth. The availability of even just motor mounts for a given mill or lathe can save a large amount of time and grief. If you don't have a huge desire to DIY everything, or if you just don't have much access to machining tools and are finding it difficult to make a certain part, then this may be the ticket to clear a major hurdle.

11

Designing a CNC Router Table

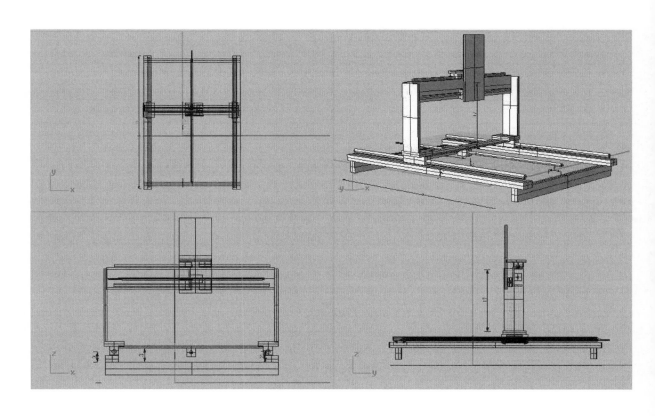

There is a lot of interest in the design of CNC router tables, large and small, for different tasks. The focus here is on the design of router tables as opposed to mills and lathes because this style of machine is more often built from the ground up by hobbyists, whereas mills and lathes are commonly purchased as pre-built manual machines, and then converted to CNC operation. Conversions can require design and fabrication as with the Proxxon conversion in the preceding chapter, but usually this is on a more limited scale than with a scratch-built router table, since the overall structure of the machine is already assembled.

In keeping with an overall goal of this book to provide a conceptual plan for creating and operating a CNC machine, I will offer you some basic, tips, information, and strategies. It is difficult to give exact, detailed directions that will be useful to a wide audience. This is true for a number of reasons, including availability of parts, differences in the tools available, and different goals. If you want exact plans for a specific type of machine which details everything down to part numbers, there are many good choices available, and some of them are listed in the references section at the back of the book. Regardless of what type of machine you want to design and build, it is highly likely that there will be some related design or construction issue covered in this overview.

CNC Router Tables: Different Design Configurations

There are some common designs for CNC router tables, and I'm going to show the more popular ones here. All of these designs have been used successfully although each has different strengths and weaknesses.

Moving Table Design

In this design, the material to be routed is affixed to a bed that is designed to travel back and forth along the X-axis. A stationary bridge above the bed holds the Y and Z axes. This design is similar to a mill in the sense that the work piece is being moved relative to the cutting tool, albeit only along one axis instead of two for a mill. The four bearing blocks for the moving table are positioned at or near the corners of the table, and the lead screw or ball screw is connected somewhere near the center of the moving table.

The disadvantage of this design is that a lot of X-axis travel is lost due to the width of the table. For example, to get a length of travel of four feet with a moving table that is two feet wide, an X-axis screw that is a little bit longer than six feet would be needed, because the table takes up two of the six feet of potential travel.

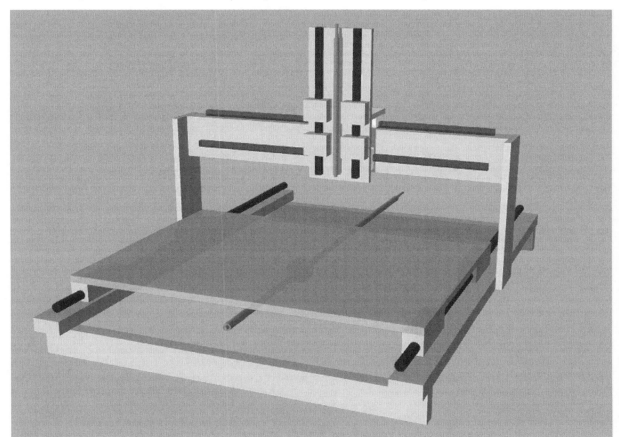

Figure 33: *A moving table CNC router table design. The stock material is fastened to the table, which is driven back and forth along the X-axis, similar to how a mill table moves. The bridge across the table holds the Y and Z axes, and is fixed to the frame at approximately the midpoint of the X-axis. The advantage of this design is that it is relatively easy to create a rigid table, which is not susceptible to twisting forces as with a gantry. The disadvantage is that the X-axis travel is significantly less than in a gantry design of the same size, because the width of the table limits the available travel. Also, a longer screw and rails will be required on the X-axis, adding expense.*

Unfortunately, longer screws and linear slides will be more expensive than with a more compact design such as a gantry (described below). Further, the extra length required in this design may result in a relatively large machine that may take up more space than desired. Given these drawbacks, the advantage of this design is that it is relatively easy to 'get right' in terms of rigidity and stability.

Moving Gantry Design

In this design, the gantry (essentially a frame that travels along the X-axis) moves relative to the bed of the router table. The gantry holds both the Y and Z axes (similar to the bridge structure in the moving bed design). The advantage of the gantry design is that the X-axis may be considerably shorter than in a moving bed design for the same length of travel. This means a more compact table that should be less expensive to build.

As mentioned before, the disadvantage of this design relative to the moving bed design is greater difficulty in designing a rigid structure than with the moving table design. This design is susceptible to different twisting forces if not properly done. This problem may be addressed by either widening the gantry, or possibly by using an additional screw along the X-axis.

Screw Configurations

There is some variation in how the screws in a router table may be configured. The major difference in these designs typically relates to how the X-axis screw is configured. Single and double screw designs have been developed.

Single Screw down the Center

As shown in Figure 34, a screw may be run down the middle of the frame, centered between the two linear rails (or rods as in this case). This design has the advantage of simplicity, and it requires only one screw to move the X-axis, making it more economical than a two screw design. The drawback of this design is that it is prone to *racking forces* when used in a gantry design. Racking refers to a twisting that is exerted on the gantry during cuts (particularly off center cuts moving along the X-axis as is shown in Figure 35) that can cause the machine to bind up during operation. Single screw gantry designs in particular require that this issue be carefully addressed during the design stage.

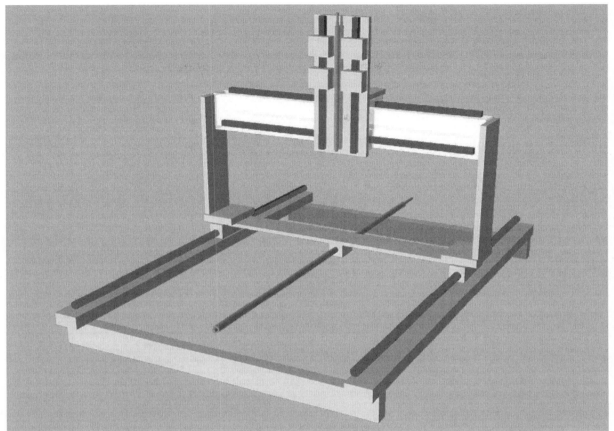

Figure 34: *A gantry design. The gantry is the frame that moves along the X-axis (the long axis) and holds the Y and Z axes. Several features (bearing blocks, motors, Z-axis platform, work surface etc.) have been excluded for clarity. These are the conceptual plans for the completed router in the case study described later in this chapter.*

Gantry designs (whether they use one screw or two) such as this one is prone to flex towards the front and back when making cuts along the X-axis. The standard way of counteracting twisting forces in a gantry design is to make sure the gantry is sufficiently wide to resist these twisting forces. The further apart the gantry bearings are, the more they will resist twisting relative to both the Y and the Z axes. The tradeoff here is that making the gantry wider will mean a decrease in the amount of available X-axis travel. It may be well worth losing a couple of inches of X-axis travel to gain substantial rigidity, though.

How Wide Should the Gantry Be?

Note: The width that is being referred to here is not the distance separating the two X-axis rails, but the distance that separates the front and back bearings on the same X-axis rail.

For a rigid design (sufficient for cutting wood or plastic router and perhaps even harder materials), the width of the gantry should be roughly half the height of the gantry as measured from the gantry bearings to the center of the Y-axis bearings. As you approach a gantry width equivalent to the height of the gantry you are getting into the realm of overkill.

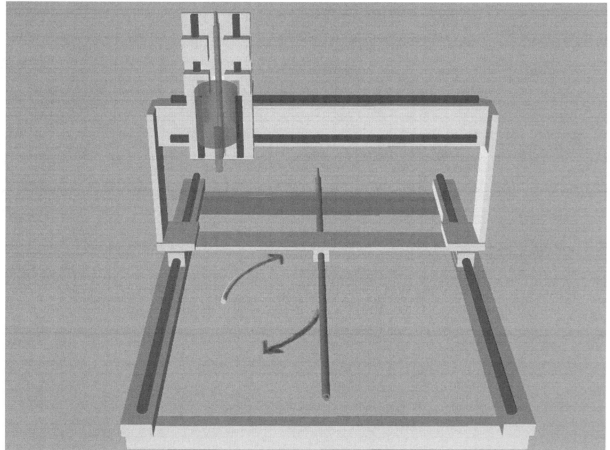

Figure 35: *Off center cuts can cause racking (twisting) that can create problems if not designed for. As shown in this figure, if the gantry is moving forward (towards the reader) and making a cut, the twisting forces indicated by the arrows would be created. A cut in the other direction (away from the reader) would result in a counter-clockwise twist. The moving table design has*

little problem with this because the inherently wide spacing of the bearings eliminates susceptibility to twisting forces.

Gantry designs that encounter relatively light forces during operation, however, may require far less rigidity than the designs described in the previous paragraph, and may get away with a narrower gantry, and overall lighter duty construction.

Double Screw Configuration

There are also designs that use two screws along the X-axis. The screws are placed in close proximity to the X-axis rails and run in parallel to them. This design has the advantage of eliminating racking forces around the Z-axis, but has the drawback of requiring two screws (double the expense of a single screw). Each screw may have its own motor, and the motors run in synchrony.

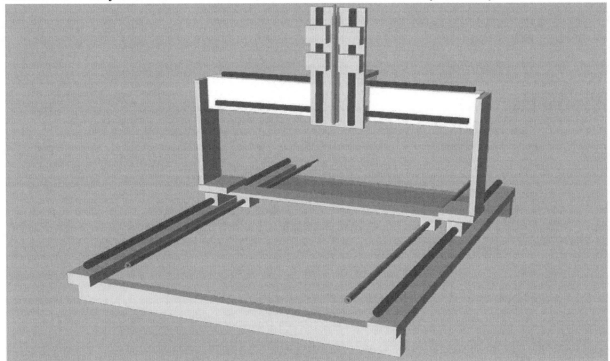

Figure 36: *A two lead screws design. The dual screws for the X-axis are located out near the X- axis rails. This eliminates the possibility of problems with the twisting forces generated by off center cuts. This comes at the expense and complexity of an additional screw and the components necessary to drive it, however.*

Although it is probably fine to drive both screws from a single motor using belts and pulleys, the author has not encountered many designs configured this way. The use of dual motors may allow for designs that are higher torque, due to the doubling up of motors. It may also allow the use of smaller motors if a single, large motor is difficult to find or too expensive.

Other Ways to Drive an Axis

Screws are a common way of driving an axis, but there are other mechanisms to do so as well. Two possibilities are a rack and pinion system or a belt drive system. These options may provide a more cost effective way of producing a relatively large motion controlled device, because they tend to be less expensive for a given amount of travel than a good quality screw.

Rack and Pinion

Rack and pinion is the same type of mechanism commonly used for car steering mechanisms, and uses a gear (the pinion) which is driven along a toothed bar (the rack) to generate travel. Rack and pinion can be less expensive than a screw-drive because the parts are relatively inexpensive when compared to a decent quality screw and nut of comparable length. Additionally racks may be laid end to end to enable long travel systems.

Rack and pinion systems are generally not as high a precision as a quality screw drive system, though, and may not be quite as smooth. This should be kept in mind when considering them for a particular application. The references section includes a source for a rack and pinion based router table plan, as well a supplier of parts.

Belt Drive System

As you may have probably already observed, belts are used as drive systems in many devices. Common examples include a belt-driven garage door opener, or an ink jet printer, which uses belts to quickly position the print heads. The basic idea in a belt drive system is to use a toothed belt and a CNC controlled motor to drive a pulley to control movement of the belt. The belt is attached at some point to the gantry or whatever structure it is being used to position.

Belt driven systems are capable of very high speeds, and a belt is inexpensive when compared to a screw, especially when long travel is necessary. They are also

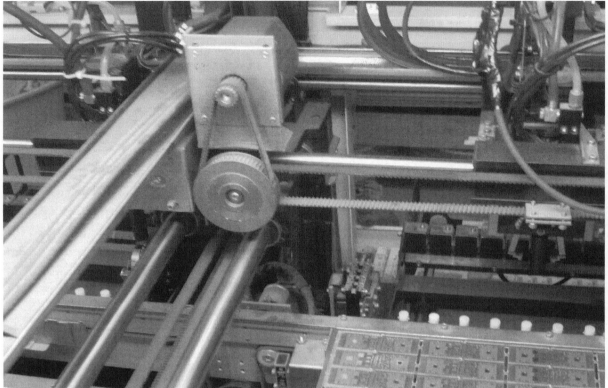

Figure 37: *A pick and place machine. It's a bit difficult to see, but this machine uses belts and pulleys to rapidly place small parts like resistors and capacitors on circuit boards. Also worth noting is the long stretches of unsupported rod. It is acceptable in this case because very little load (weight or force) is placed on the rods during operation, because the machine is only lifting and placing small, light items, not moving a heavy gantry and making cuts into hard materials.*

relatively quiet during operation when compared to a screw. However the lack of rigidity (compared to something like a screw) may limit their applicability to situations that don't require a lot of rigidity. This tends to exclude belts from being used for an application like high precision routing or milling of hard materials. They are ideal in some situations, though.

A pick and place machine is one such example (Figure 37). These machines are used to automatically load electronic components on a circuit board, and are capable of amazingly fast operation over relatively long distances. Belts are great for long travel, high speed applications that do not require that a lot of force be generated to move heavy loads or make difficult cuts in a precise fashion.

There are plans for belt design machines available, as well as examples floating around the Internet, so if you think this might be a good option for your particular application, look in these places. You can also examine familiar devices (the aforementioned ink jet printer or belt drive garage door opener) to see how a belt drive system is implemented in different situations. Manufacturers of belts provide information on their websites about the use of their products for different applications.

Rotating Nut Designs

This is a relatively uncommon design where the nut is spun by a CNC controlled motor around a stationary screw. The nut is driven along the screw as it spins, and drives whatever it is attached to, such as a gantry or other structure. This type of system has some interesting benefits:

1. Since the screw does not rotate the screw may be held under tension, allowing for the use of narrower (and cheaper) screws than normal.
2. The lack of screw rotation helps eliminates the problem of whipping by the screw.
3. The motor does not have to spin a heavy screw up to speed.

For DIY and hobbyists this is an unusual design choice, and requires manufacturing a bearing system to hold the nut and another mechanism to spin the nut with a motor. Although this is an interesting design, it is probably easier to stick to the other drive systems mentioned, because it would be difficult to create this type of design without substantial access to an existing CNC machine.

Designing a Router Table - A Case Study

The following is a discussion of the development and construction of a CNC router designed for cutting wood and plastic. I'm not suggesting that you build a machine like the one presented. You may have something entirely different in mind; smaller, larger, or designed to cut different materials. It provides a useful platform for discussing basic design features and choices you may be faced with when designing and building a CNC machine yourself.

This machine was built with a very limited selection of tools (a drill press, hack saw, and a few others), and a small amount of outside assistance. It is also a fairly generic style of router table, and it is truly a DIY machine. Some aspects of it are not exactly pretty, but it works reliably for its intended purpose.

Construction Materials

The CNC router table (pictured in Figures 39-43) is constructed almost entirely of aluminum. The gantry is made from precision cast aluminum plate, and most of the frame is constructed from aluminum extrusion known as 80/20 style extrusion, T-slot extrusion, or modular aluminum. This style of extrusion is versatile, and may be cut to length and bolted together like an erector set. It also has other advantages; it is possible to position and bolt almost anything easily along the length of the aluminum channel, and reposition it if necessary without repeatedly drilling and tapping holes. It is also easy to run wires within the slots that run along the length of the channel, making for cleaner wiring. The precision ground aluminum plate has the advantage that it is extremely flat on its surfaces.

The drawback of using a lot of aluminum is that aluminum has become significantly more expensive over the past couple of years. This is probably the main restriction in using this material because in general, it is a very convenient way to construct a frame for a CNC machine.

Aluminum extrusion comes in a variety of sizes and there are a variety of brackets (right angle brackets, plates with holes drilled in them, etc) available to allow assembly into all sorts of configurations. It may be *disassembled* just as easily as it may be assembled, in case an error is made and an adjustment is necessary. Manufacturer's web sites have plenty of information about their products, and how they are used. As mentioned before, aluminum channel is expensive, so it is advisable to scout around a bit to see if some of this material is available at a recycler if it is going to be used extensively in your project. An Internet search should result in a list of at least a few individuals who sell surplus T-slot extrusion.

Other Materials

Another candidate for building a frame is square steel tubing. If you can weld well or know someone who can, it is possible to fabricate a relatively inexpensive and rigid frame from this material. This kind of frame may be used to support a flat surface to which linear rods or rails may be attached. It is also possible to use steel pipe (available at most hardware and home improvement stores) to assemble a frame for small machines relatively quickly and cheaply.

Figure 38: *Four pieces of "T-slot" aluminum extrusion and a piece of cast aluminum plate (with the protective plastic peeled back to show the surface). The T-slot may be cut to various lengths and bolted together to form a framework for a CNC machine. Precision cast aluminum plate is extremely flat; a useful property for making precision machinery. Unfortunately, in recent years aluminum has increased substantially in price.*

There are some recent hobbyist designs that use significant quantities of MDF in their construction. MDF is a high quality particle board style material that is very flat and consistent, and is available at home improvement centers. In these designs, identical ribs are cut from MDF and are used to form long, relatively rigid structures to use as a base and gantry. The goal is to avoid the high cost of aluminum or steel, and it appears that some good results have been achieved. Usually these machines are intended for cutting wood and plastic. Designs are available on the Internet, and there are several examples of completed machines.

The Basic Design
The design of this table is a relatively standard screw down the middle moving gantry design. The gantry style was chosen because it allowed for a relatively compact table with substantial X-axis travel. This compactness also resulted in

cost savings because the extrusion length, linear rails, and drive screw for the X-axis were all shorter than they would be for a moving table design with equivalent X-axis travel. A second X-axis screw was considered but excluded because of the added cost and complexity.

Figure 39: *At an early stage of construction. The basic framework is of four pieces of extrusion bolted together at right angles, with one inch diameter precision round rods bolted to them. The gantry was built from cast aluminum plate and a piece of C-shaped aluminum extrusion; it is essentially a rectangular frame that is attached to the bearing blocks. The Y-axis was created by bolting two linear rails to the extrusion. A piece of extrusion is placed in front of the gantry to keep it from rolling off the end of the table, because there is nothing to prevent this. The table is approximately 56 inches long by 40 inches wide.*

This design first took shape by purchasing the X-axis components (round rod, supports, and bearings) and the aluminum extrusion to build the frame. A CAD program was used to model possible designs based on these starting pieces.

The Drive Mechanism

Both ball screws and lead screws were considered for this design. Ball screws tend to be very efficient (they lose little energy due to friction) so may not require as much motor power to move a given load. Lead screws have the advantage that they are less vulnerable to the dust and debris produced by a router table (the nut is solid and doesn't accumulate crud like a ball nut does). Since this machine is primarily used for cutting wood, lead screws would be a better choice from a maintenance standpoint, and probably wouldn't have given up too much in terms of efficiency (some modern lead screws are quite efficient). Initially it was decided that lead screws would be used for this reason.

Ultimately, however, ball screws were chosen. They were purchased in part because they were close to the desired length for the initial design, were of a decent diameter, and came with machined ends. As mentioned elsewhere, machining the ends of screws (to fit into bearings) is difficult, and in most cases must be done by someone with proper experience and equipment. In this case, it was just cheaper to purchase screws with ends machined at the factory than it was to purchase lead screw material and hire someone to machine it. If suitable lead screws with machined ends had come available, then this machine would probably have lead screws instead of ball screws. It was more a matter of opportunity.

Note: For small, lightweight designs it is possible (and not uncommon) to use threaded rod with locking nuts or collars with a bushing or bearing to get around the need for machined ends. For tables with long travel requiring fairly long and large diameter screws like this one it would be difficult to create a successful design this way.

Bearings and Bearing Blocks

The screws for the X and Y axes are supported in bearings at both ends. The blocks that hold the bearings are known (sensibly) as bearing blocks. The bearings perform two functions; the primary one is to allow the screw to rotate freely and smoothly. They also help hold the screw in place, and must resist the forces that are transferred back along the screw during operation. Figure 11 gives an illustration of these forces.

For the DIY builder, it may be e a good idea to engineer some adjustability into the placement of bearing blocks. The bearing blocks in this design were fashioned out of plastic using Forstner bits because the builder didn't have access to milling equipment to manufacture this part. Plastic is a perfectly adequate material for holding a bearing, and is easier to machine than aluminum. In this case, the T-slot extrusion provided a lot for side to side adjustability, and slightly oversized bolt holes provided a little more up and down and side-to-side adjustability.

Figure 40: *A close-up view of the back side of the Y-axis, showing the ball screw, plastic bearing block, pulleys, and belt. The belt runs from the motor to the large pulley. A smaller pulley (the black one in the lower right corner) touches the belt midway between the motor and the large pulley, and is used to turn the shaft of a panel mount encoder (see Figure 23) for feedback. It would be better to have the encoder mounted to the back of the motor, but this arrangement has some advantages which are detailed in the text.*

There are several bearing suppliers online who have been around for awhile and have good variety and prices, so I would recommend investigating these sources, especially if you need an uncommon diameter. Getting items such as bearings, belts, and pulleys from a local supply may prove expensive.

The motors chosen for this machine are used servo motors sourced from one of the surplus houses listed in the reference section at the end of this book. These particular servo motors were chosen for a number of reasons such as cost, availability, and a need to get started. They have performed well for this application. Either stepper or servo motors could be used successfully in this design, however. There are some operating characteristics of each type which may indicate you should use one over the other in your own project, as described in Chapter 5.

Motors and Encoders

The motors used in this machine have a peak torque of approximately 400 oz inch, are very durable and smooth, and were obtained second hand for far less than new servo motors with comparable ratings. The disadvantage of these came in the form of a short shaft and no encoders (the center motor in Figure 6). An initial attempt was made to modify the motor by adding an extension to the back end of the motor shaft to allow addition of a through-hole encoder. This proved to be too difficult given available resources. The encoder issue was ultimately solved by mounting the encoder along the path of the belt in the belt and pulley system that connects the motor to the ball screw. It would be simpler to have direct mount encoders (like the motors on the left and the right in Figure 6), but the motors themselves were not originally created for this purpose.

There are some drawbacks to this particular arrangement. The encoder used in this case is a panel-mount encoder (Figure 23, right hand side), and is not really designed for this use so may not wear particularly well. The mounting technique is not ideal, because the shaft is not supported at both ends, and this puts a load on the side of the encoder shaft. This system has worked relatively well, however, but will probably require that the encoders be replaced more frequently than desirable. This wouldn't really be good for a situation where the machine is heavily used, but is OK for a more moderate duty situation. These choices were made for the sake of using inexpensive, but high quality, used servo motors.

Belt and Pulley

A belt and pulley system was chosen to connect the motor to the ball screws for a several reasons. The screw is metric while the motor shaft is not, and this meant that a commercial coupler would extremely difficult to find, and very expensive regardless. The manufacture of such a coupler would have been difficult or impossible given available resources at the time. Further, it's relatively easy to find individual pulleys to fit almost any diameter shaft.

Additionally, a pulley and belt system is also easier to align than a direct drive design. A further benefit is that the gear ratio may be changed by changing the pulley on the screw and the pulley on the motor. This allows the designer to tailor performance (exchanging top speed for maximum force or vice-versa) merely by changing the relative diameters of the pulleys, as described in Chapter 4.

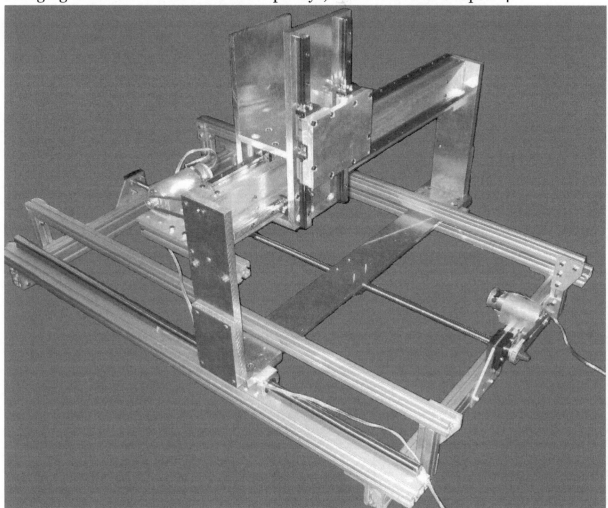

Figure 41: *Further progress on the table includes the addition of the Z-axis rails and platform, and mounting of the X-axis and Y-axis ball screws. The Y-axis ball screw is not visible in the picture, but is shown in Figure 40. Motor mounts and motors have been installed for the X and Y axes, but not yet for the Z-axis. Two long pieces of aluminum channel have been put in place as a support structure for the surface of the table, and smaller pieces will connect across the two longer pieces like the rungs on a ladder to prevent sag. The ball*

screws are supported by bearings at each end using plastic bearing blocks, which were designed to allow some adjustability in placement.

The drawback of this system is some increase in complexity and probably less precision than with a direct drive system, but it is difficult to determine by how much.

Linear Slide Mechanism

The linear slides used in this design are a combination of precision round rod with matching bushings, and linear rail and bearing blocks (Figures 12-14 provide examples of these devices). The round rod is used on the X-axis, and linear rail is used on the Y and Z axes. These devices were chosen largely because they were available as the time and were the right size.

Linear slides must be mounted to flat surfaces. Certain materials that appear flat may not actually be, and some warp in an apparently flat surface may exist without being obvious to the naked eye. Linear slides attached to surfaces will most likely to conform to the surface they are attached to, flat or otherwise. The surfaces that the Y and Z-axis linear rails were attached to were checked for flatness prior to mounting by holding them up to a piece of precision ground aluminum plate, which has a surface that is known to be very flat.

Please note that the aluminum extrusions used to make the frame have slightly canted (i.e. angled slightly upwards, like a roof) sides which is a deliberate design feature. They may not be suitable for direct attachment of some types of linear slides. The round rail used in this case is suitable for direct mounting to the T-slot extrusion, because this style of rail has some rotational play that can compensate for this.

The Electronics

The electronics in this system consist of a PC (running Windows XP and using Mach 2 as a controller program), and a controller box. The controller box was created using three servo drives, a transformer, capacitor, basic breakout board, and a few other components like switches and terminal blocks (see Figure 19). The components were chosen because of availability and adequate performance characteristics. There are other controller programs, servo drives, breakout boards, and other parts that would probably work as well or better; it's just a matter of doing a bit of research to make sure the components are compatible.

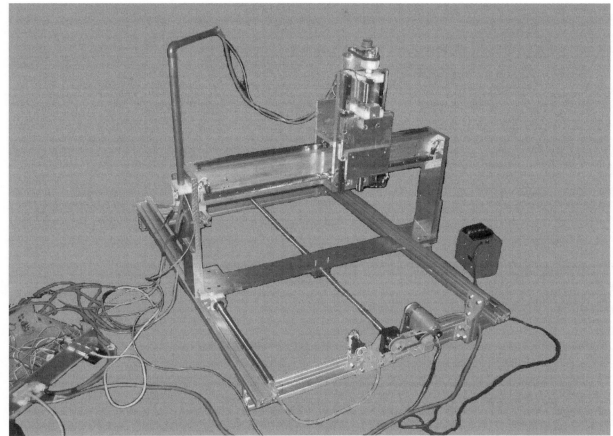

Figure 42: *The router table nearing completion. All screws, motors, encoders, belt drives, wiring have been installed, and the motors and motor controller box are being tested. The controller box shown at left is the same one shown in Figure 19, and receives control signals from a PC (not shown) and gets feedback from the encoders on the router table. The copper tube that sticks up at the side of the machine is to help keep wiring from snagging during operation, a common problem when making CNC machines with long travel on an axis. There are better ways to deal with this problem (see Figures 43 and 46).*

Dust Issues

It is smart to think about dust collection when designing a CNC machine. This is especially true for machines that are intended to cut wood, MDF, fiberglass, circuit board material and other substances that turn into powder when milled. This dust can and will get everywhere. This includes on the linear slides (and into the bearing blocks), into ventilation systems, and into your lungs. This is potentially a very serious health issue which should not be ignored.

The best solution is to have a vacuum system that grabs most or all of the dust as produced, but in practice this is difficult to achieve. For the dust that does escape a vacuum system, it is good to design a machine so that parts with ball bearings, such as ball screws or linear slides are shielded from dust (or enclosed if possible).

In this design, the long X-axis screw and the X-axis rails are placed below the cutting surface which keeps dust from directly falling on them. The Y-axis ball screw is recessed inside the aluminum extrusion that runs across the top of the gantry (see Figure 40), so it is well protected from direct exposure. The Y and Z-axis rails and the Z-axis ball screw (on the gantry) are above the cut surface so dust will not drop directly onto them, but unfortunately they are still exposed to any dust that billows up during cutting. The better protected mechanisms containing ball bearings are from dust, the longer they should last.

With small mills that are primarily used for cutting metal and plastic, dust will not be as big a concern as these materials tend to produce fairly large chips that don't take to the air readily. They still produce a mess, so a containment mechanism might be a consideration, but it probably shouldn't be as big an issue as with the situation described above. With a small machine, it is possible to take a brute force approach, and fully enclose the machine in a box or tent of some sort.

Discussion

Many design decisions during this build were forced by a prior decision. The choice of the motors was based on cost considerations, and the choice of motor drive was based on the choice of motor. The choice of a pulley and belt drive system was motivated by the potential difficulty in manufacturing a suitable coupler. This choice resolved a problem in using encoders, and it offered flexibility in setting up the operating parameters, such as the resolution and speed of the machine. Things progressed in a somewhat domino like fashion after the initial design choices were made. This is not to say that the design of this machine was a haphazard process; far from it. There was a significant amount of modeling and design that went into its creation. However, the use of surplus parts, and the limited amount of machining available required periodic reevaluation of the design along the way.

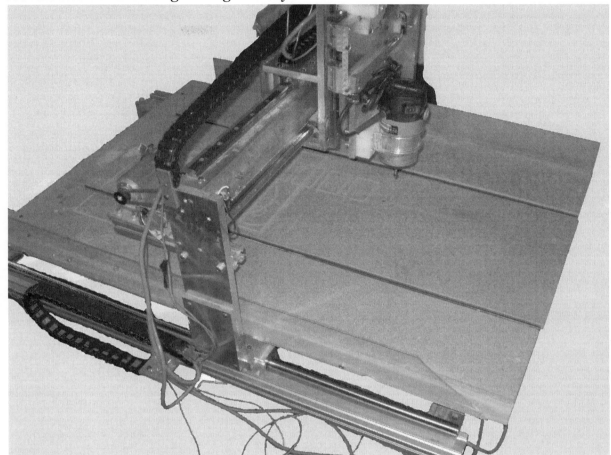

Figure 43: *The finished router table with 1¾ HP router for a spindle. The copper tubes that guided the wires have been replaced with a much tidier cable guide system. The surface is particleboard, which is relatively flat, inexpensive, and can be cut into repeatedly and replaced when necessary (if you look closely you can see some circles, rectangles and other shapes that have been cut down into the table). MDF would be a better choice as it is even flatter and more uniform than particle board.*

12

Design Calculations

Some basic calculations are necessary when designing a CNC machine or converting an existing machine. A common problem, for instance, is how to calculate the maximum travel that might be achieved with a particular setup (combination of motor, screw, motor drives, and controller program settings). The math required for much of this isn't particularly difficult; just basic algebra.

Calculating the Lead of a Screw

The lead of a screw is the axial (straight-line) distance the nut (and whatever the nut is driving, such as a gantry) will move with a single turn, and is calculated by determining the distance between adjacent peaks and then multiplying by the number of starts of the screw. For instance, a piece of threaded rod from the hardware store that is listed as 1/4 − 20 is a quarter inch in diameter with 20 threads per inch), and would have a distance of .05 inches between each thread, since 1/20 = .05. Threaded rod has a single start, so the distance that the nut would move for a single turn of the screw is .05 inches.

If you have a lead screw with a distance between the threads of .125 inches, and it has four starts (the screw at the right in Figure 9 has these specifications) it will have a lead of .125 x 4 = .5 inches. One rotation of this screw will move the nut half an inch. If you are unsure of the lead of a given screw, you may get an idea by rotating the screw several times then measuring the distance traveled by the nut, and then dividing by the number of rotations.

Calculating Speed of Travel

Calculating the speed of travel for a particular motor and screw combination requires that you know the rotational speed of the screw and the lead of the screw. Specifically:

Speed of travel = lead x rotational speed

For example, consider a screw that has a lead of .1 inches. If a motor is directly driving it and is rotating at 500 RPM, then the nut will be pushed along at .1 * 500 = 50 inches per minute, as will whatever the nut is attached to. Typically these calculations are done to determine the maximum speed of travel (also known as the *rapid* speed). If the motor has a top speed of 2000 RPM, the top speed of travel in this case is .1 x 2000 or 200 inches per minute.

There are a number of factors that may affect this speed. Even if a motor is listed as being capable of a particular top speed, in practice this speed may be less. A stepper motor, for example, may be able to spin at high speed but it may not have adequate torque at high speed, hence the actual (perhaps *practical* is a better word) top speed is less. Other factors that might influence this top speed are the quality of the electronics, such as the motor drives. Some motor drives perform better than others and can generate higher speeds and higher torque for the same motors.

Calculating Resolution

Resolution refers to how small a move a CNC machine can reliably make, and it affects both the precision and accuracy of the machine. It is a function of the lead of a screw and the number of steps that make up a revolution using a specific combination of motors and motor drives. In the case of stepper motors, the resolution is calculated as the lead of the screw divided by the steps per revolution for the motor multiplied by the number of microsteps (if any) the motor drive is generating. For instance, a stepper motor with 200 steps per revolution operating in quarter step mode (i.e. each step is subdivided into four microsteps) would have 800 total steps per revolution. If this motor/motor drive combination is used with a screw with a lead of .2 inches, the theoretical resolution would be .2/800 = .00025 inches. This means that the CNC machine could (in theory) make movements as small as .00025 inches.

In the case of a servo motor, the resolution of the motors is determined by the resolution of the encoders being used. For example, 128 CPR encoders operating in quadrature will produce 512 pulses (4 * 128) per revolution of the motor. This motor combined with a lead screw or ball screw with a lead of .2 inches per revolution would have a theoretical resolution of .2 / 512 = .000390625 inches. That is to say, the smallest distance that this setup could move is .000390625 inches.

The reason the term theoretical resolution is being used here is because several factors can (and usually do) change the true resolution of any CNC setup. Factors such as screw non-linearity, backlash, and uneven microstepping typically reduce the actual resolution of the machine. This, however, represents a reasonable first approximation, and can be taken as an upper limit of the machine's capability.

The Downside of High Resolution

Extremely high resolution may sound good, but can limit the top speed at which your machine can move. This is because there is often a limit to how fast steps may be generated in some circumstances (one reason is described in the next few paragraphs). A shorter distance will be traveled when taking small (i.e. high resolution) steps. All else being held equal, the top speed in the high resolution machine will be lower than in a low resolution machine because the lower resolution machine will take bigger steps each time.

Complicating Things a Bit

There are a couple of other factors that can complicate the calculation of the speed capable in a given machine; one relates to a technical issue with the use of the parallel port, and another relates to the presence of gearing between the screw and motor.

The speed at which the parallel port can send step signals depends upon a number of factors, including the speed of the computer and the software being used. Ultimately this can limit the top speed of a machine that is using the parallel port to send step and direction signals to the motor drives. For instance, if the parallel port on a particular computer can produce step signals at 50,000 Hz (cycles per second), and at this rate the machine has a top speed of 150 inches per minute, switching to a computer that can only produce step signals at 25,000 Hz will cut the top speed to 75 inches per minute.

As first described in Chapter 4, the gearing between the motor and the screw can also influence the speed of travel. For example, a gear on the screw that is twice the diameter of the gear on the motor shaft driving it will result in a halving of the rotational speed of the screw, cutting the top speed in half. This will also double the torque at the drive screw, will double the force it is capable of generating, and will double the resolution as well.

Tradeoffs: Resolution, Speed, and Force

The resolution of a machine refers to its ability to move a small increment, and force refers to the maximum amount of 'push' that the machine may generate along a given axis without stalling. High resolution, high speed and high force are all desirable traits, because they allow for easy and very precise cutting through even very hard materials. If it were possible to have all three on a CNC machine without spending a fortune, then life would be easy.

Unfortunately, real life requires that compromises be made, and in the case of CNC machine design, the trade-off between these three factors is one of the most fundamental. The following table lists different ways a CNC machine may be modified, and how it impacts aspects of machine performance.

How Design Choices Affect Maximum Speed, Resolution, and Force

Design Choice	Speed	Resolution	Force
Increase lead of screw	increases	decreases	decreases
Decrease lead of screw	decreases	increases	increases
Gear motor down	decreases	increases	increases
Gear motor up	increases	decreases	decreases
Increase encoder resolution	decreases	increases	no change
Decrease encoder resolution	increases	decreases	no change

This table assumes that nothing else is changing other than what is specified under the *Design Choice* column. If for instance, the calculated top speed is less than desired, it might be possible to increase speed by the use of gearing between the motor and the lead screw or ball screw it is driving. This would come at the cost of decreased resolution and decreased force if nothing else were changed. You could obviously change a combination of these things to get a desired result as well. Increasing the lead of the screw and decreasing the encoder resolution, for instance, would produce a faster top speed since both of these choices will increase speed individually. This would also decrease the resolution of the machine and decrease the maximum amount of force generated too, because these changes will produce these effects as well.

13

Wiring

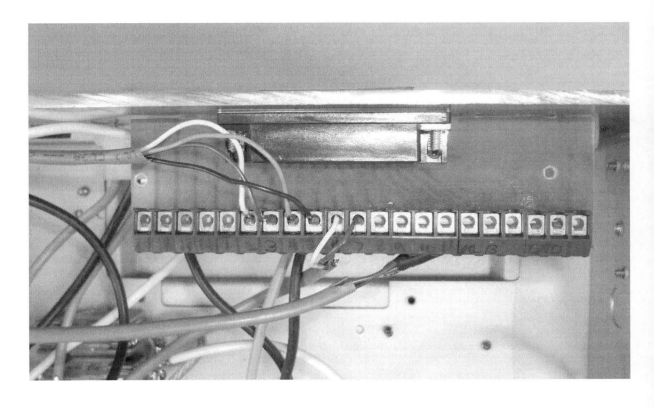

CNC machines require a fair bit of wiring. Motors, limit switches, home switches, encoders, and other devices all require runs of wire, and oftentimes require several leads per device. This section provides some advice about layouts, what materials to use, and where to find them.

Wiring of Motors

There are a few basic considerations when wiring a CNC machine. These include determining which gauge of wire is sufficient to handle the current demands of the motor during operation, how to run the wire so that there isn't clutter or potential snagging during operation, how to properly wire home switches, limit switches and other devices, and how to wire motors, encoders and such to avoid possible electrical interference.

Wiring Stepper Motors and Servo Motors

Stepper motors require that at least four independent wires be run from the motor drive to the stepper motor. For the sake of tidiness and sanity, it is smart to bundle the wires together to form a single cable. Fortunately multi-lead cable may be purchased cut to length at a home improvement center. Some very useful stuff is referred to as shielded, twisted pair and may be found with the other spools of cut-to-length wire products such as a lamp cord, bell wire, stereo wire, etc. This is available with different numbers of leads in a single cable. An advantage of this product (beyond the tidy packaging) is that the wires are shielded with foil, which can reduce electrical interference. Correctly wiring a twisted pair (see Figure 44) may help to reduce electrical noise during operation.

Wiring of Encoders

As with stepper motors, multiple wire leads are required to connect an encoder to a motor drive. Quadrature encoders typically require at least four wires (one wire for each of the two out of phase pulse streams, one for ground, and one as a voltage supply for the encoder). The two encoder channels are typically referred to as A and B, and the ground and voltage supply are commonly referred to as + and −, or V_{cc} and GRN (for ground) or something similar. The easiest way to deal with this potential wiring mess is to use the multi-lead cable described above. The shielding of the individual leads provides some additional insurance against interference that could lead to operational problems.

It is advisable to separate encoder lines from the motor drive lines (i.e. the wires carrying current for the servo motors) as best as possible, because they might cause interference in the encoder lines.

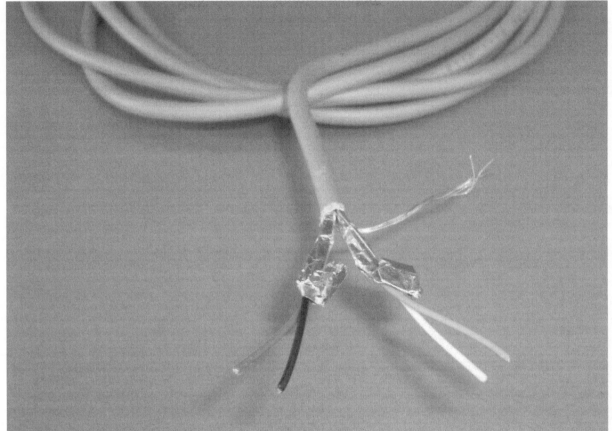

Figure 44: *Shielded, twisted pair cable is useful for wiring motors, encoders, and so forth because of the large number of leads in a single, flexible cable. The shielding may be useful in eliminating potential electrical interference. The bare wire sticking out to the right may be wired to act as an antenna, to help reduce outside electrical interference.*

Twisted Pair

A twisted pair is a pair of wires which are twisted together along their length to help reduce noise and crosstalk. The most beneficial way to wire a twisted pair depends on the application. With a stepper motor, for instance, a twisted pair should be wired with one half of the pair carrying the current out to the device and the other carrying the return of the same current back. When wiring a four-wire (bipolar) stepper motor with a cable containing two twisted pairs, one pair is dedicated to one of the motor coils, and the other pair is dedicated to the other. When using twisted pair with encoders, it is best to avoid putting both encoder channels (A and B) on the same pair, to avoid inducing noise in the other channel. These are merely suggestions, and violating them may or may not have an impact on performance. Being aware of this may come in handy when troubleshooting, though.

Switches and Proximity Sensors

A variety of different switching devices are used to indicate if a CNC machine has reached a particular location, such as with home switches and limit switches. Home and limit switches may be physical switches, where a circuit goes from open to closed when a lever is depressed, or more exotic devices like a proximity detector or optoelectric (light sensing) device which doesn't rely on physical contact. Proximity sensors operate through detection of small changes in inductance or capacitance caused by the presence of an object near the sensor. Inductively triggered proximity sensors detect metal objects, and capacitance triggered proximity sensors are triggered by non-metal objects.

Ideally, the switches used for home switches should exhibit little variability in the position at which they trigger. If the switches used are somewhat sloppy and don't always switch at the same position, then they can make repeatability in a production run difficult as well as complicating machine setup.

For example, if a part is being cut and for some reason the controller software has problems (glitches, crashes, or the power to the computer is cut off), the controller program will lose track of the current position of the CNC machine. The controller program doesn't truly know where the machine is positioned; it just knows how far it has told the motors to move relative to where it thinks home is. Home switches are designed to provide an absolute reference (a specific, physical home position located in the travel area of the machine) which is independent of the 'memory' of the controller program, that can be located if for some reason the true position is lost.

Switches that don't trigger reliably, however, will not provide an accurate enough reference for a successful recovery. They may get 'pretty close', but for many machining operations, this might not be good enough. This will be true for other situations where an absolute reference is desirable, such as when the operator wants to stop in the middle of a long run until the next day.

If a switch is intended to be used strictly as a limit switch, then high precision is not as important. It should trigger prior to reaching the end of an axis to keep a machine from being damaged by running too far along an axis; and some variance in the position where they trigger probably isn't going to be a big problem.

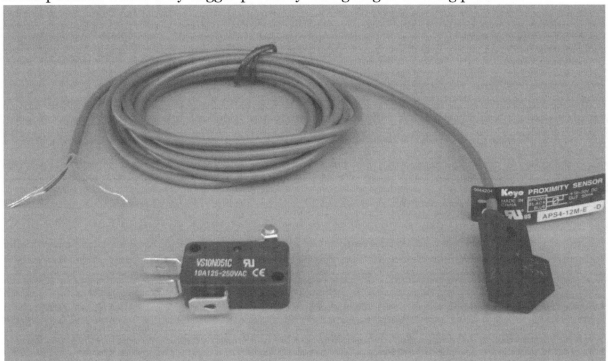

Figure 45: *Two switching devices. The device at the bottom is a micro switch, and relies on the physical depression of a lever to make it switch. This particular switch may be wired so that depression of the lever either closes (connects) an open circuit, or opens (breaks) a closed circuit. The other device is a proximity sensor which triggers when a piece of metal comes within a specific distance of the sensor.*

Wiring of Limit Switches

As described above limit switches are designed to signal the controller software to shut down the motors on a CNC machine when the normal limits of travel have

been exceeded. A wiring diagram is shown in Figure 46. The switches are wired
to an input pin and a ground pin of the parallel port. All are wired in series and
are normally closed (i.e. they are completing the circuit). If any one of the
machine's limits is exceeded, the circuit is broken, and this signals the controller
software to stop driving the motors.

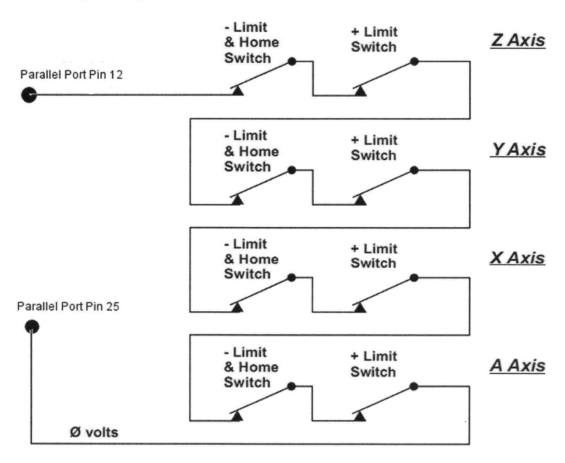

Figure 46: *Limit and home switch wiring diagram. Parallel port pins 12 and
25 are used here, but other combinations are possible, provided one is an input
pin and another is a ground pin. In this case, one set of switches double as home
switches. The A-axis is for a rotary axis, if one is present in the machine.*

It is possible to wire limit switches so that they are in line with the servo motor.
When the switch is triggered, it goes to an open state which cuts power to the
motor. This is not recommended, because it can potentially generate back EMF
(a voltage spike) that can damage the output transistors on a motor drive. This

setup might be used, however, as a failsafe to absolutely stop a runaway motor (as a backup to the limit switches). This would provide additional insurance in the case when a driver has failed and continues to drive a motor even when the software signals it to shut down. This kind of "belt and suspenders" approach might be useful in a high powered servo system.

Controlling Wire Movement in Larger Machines

In CNC machines with relatively long travel, some thought should be given to dealing with the possible interference between the moving parts of the CNC machine and the bundle of wires connected to various parts of the machine.

Figure 47: *Igus 'Energy Chain' is designed to neatly package and guide cable in machines with relatively long travel on at least one axis. It's more elegant and functional than the makeshift cable guide shown in Figure 42).*

In a gantry router with X-axis travel of a few feet or more, there will be a considerable mass (mess?) of wire running to the gantry, which must travel along with the gantry as it moves back and forth along the X-axis. Probably the best solution is to use a cable carrying and guiding system (see Figure 47). This is expensive (try purchasing it second hand if possible), but is a very slick way to deal with what can be an annoying problem. Figure 42 shows a makeshift solution cobbled together from copper pipe. It's cheap and works OK, but isn't pretty, and limits visibility somewhat.

Length of Wire

Low level signals, such as the digital signals sent from an encoder back to a motor drive, or those sent from the parallel port to the motor drives, may be degraded by the electrical resistance caused by long runs of wire. In most cases, lengths of shielded cable or parallel port cable of 20 feet or less should be OK. In the few instances where long runs of wire are desired or cannot be avoided, it is possible to use a device called a line-driver to amplify a low level signal so that it will be strong enough to deal with the resistance of a long wire lead. Some encoders have this circuitry built in.

14

Software

```cpp
#include <fstream.h>
 long flength(char *name)
 {      long length = -1 L;
      ifstream ifs;
      ifs.open(name, ios::binary);
      if (ifs)
      {
          ifs.seek(0L, ios::end) ;
          length = ifs.tellg() ;
      }
      return length;

 }
#else
 long length(char *name)
 {

      long length = -1 L;
      FILE *fptr;
      fptr = fopen(name, "rb");
      if(fptr != NULL)
      {
          fseek(fptr, 0L, SEEK_END);
          length = ftell(fptr);
          fclose(fptr);
      }
      return length;
 }
#endif /* C++ */
#ifdef TEST
```

Software

There is a bunch of software involved in CNC machining. Usually there is a CAD program, a CAM program, and a controller program in the pipeline. These programs allow you to draw a part (CAD), create instructions to control a CNC machine called G-code (CAM), and send control signals to the controller electronics based on this G-code (the controller program). Others applications may be included as well, such as a G-code editor and a program to quality check new G-code before it is run. A synopsis of each of these program types is given below:

CAD (Computer Aided Design): CAD programs are basically high precision drawing programs intended for design purposes. CAD is used for the creation of accurate diagrams and models of whatever the operator is interested in creating. Different CAD programs may be better suited for particular tasks.

CAM (Computer Aided Machining): CAM programs read files from a CAD program and create G-code based on it. G-code is a standard way of specifying a CNC machine's movements and actions. G-code is read by a controller program. CAM automates most of the production of G-code.

Controller program: A controller program reads G-code instructions and outputs a stream of signals to the motor controller box to control the movement of the motors on the machine. Essentially, the controller interprets G-code and tells the motor drives how to spin the motors in real time to cut the part described by the G-code.

G-code editor: G-code may also be written by hand with a simple text editor, such as a notepad program, or with a more sophisticated program specifically designed for editing G-code. These editors may have niceties such as colored syntax highlighting and wizards for generating code. It is possible to create complex parts coding by hand with some practice.

CAD/CAM: CAD/CAM refers to a single program that integrates both a CAD program and a CAM program or to a pair of programs used together in this way. The term CAD/CAM can also be used to refer to the whole process of designing a part and create code to manufacture it using CAD and CAM programs.

Toolpath verification software: Toolpath verification programs are designed to help a CNC operator verify if G-code is correct (will do what is expected) before actually trying it on a machine.

More about CAD Programs

Complex parts are usually created by drawing them in a CAD (Computer Aided Design) program. The files that a designer creates using a CAD program can be opened in a CAM program to generate G-code.

CAD programs vary as to their focus. Some are geared towards creating blueprint style drawings and others are more focused on 3 D modeling. Some CAD programs (or plug-ins for popular CAD packages) are highly specialized for tasks such as jewelry making or shoe design.

Different types of CAD programs are better matched to particular types of machining (see Appendix A for a detailed description). A drafting style program is more applicable for 2½ D machining purposes, and a 3 D modeling program is more applicable for 3 D and four and five axis machining. Many can do a bit of both, so one program may cover most needs. If you don't have prior experience using a CAD program, be prepared to commit some time doing so, as there is a bit of a learning curve with most of them. It's a smart idea to make simple drawings initially, and try to follow tutorials available with the package or on the Internet.

CAM Programs

Like CAD programs, CAM programs may be oriented towards a specific purpose or may be more general in nature. Expensive and sophisticated packages usually offer more types of machining capabilities, such as 2½ D, 3 D, and fourth and fifth axis capability. Some CAM programs may have fairly unique capabilities such the ability to turn picture files from a PC (such as a jpeg file) into G-code used to make a carving or engraving of that picture. If you have a specific application in mind, there may already be a CAM program designed for that purpose.

CAM programs read different types of CAD files. Common file types include DXF files for 2 ½ axis machining and STL files for 3 axis (and higher) machining. DXF is a file type originally created by Autodesk (the makers of Auto CAD) and is supported by many CAD programs. STL is a file format commonly produced by 3

D CAD programs, and contains a mathematical description of the surfaces that make up a 3 D object.

Learning to operate a CAM program is not necessarily as involved as learning a CAD program, but it does require some practice to understand how to configure things to produce useful G-code. In the case of most CAM programs, the dimensions of the stock (the raw material to be machined) must be specified, as does the machining strategy, the type of cutter used, feed rates, and so forth.

An understanding of advanced issues, such as how to optimize cutting paths and how to produce the best quality cuts and finishes comes with a fair amount of trial and error. Also, CAM programs tend to be idiosyncratic, so it can take a while to become comfortable with their individual quirks.

File Importation/Exportation Issues
It's possible for compatibility problems to occur when importing and exporting files between the different CAD and CAM programs. For instance, some programs may not be able to handle all versions of DXF files, as the file format has changed over the years. This is an issue that is addressed as it crops up, however. More robust programs are able to handle more file formats and with fewer problems. Sometimes outputting to a different version of a file type will fix the problem; for instance, an older CAM program may be able to import older versions of a file type, but not newer ones.

What is a Post Processor?
Although G-code is supposed to be a standardized language, different manufacturers of CNC equipment have over the years created somewhat different versions of G-code to work with their particular machine. A post processor formats the G-code produced by a CAM program so that it will run properly with a particular manufacturer's machine. A CAM program that says it includes "40 different posts" is indicating that it can produce 40 different "flavors" of G-code to run with different machines. There are stand alone programs that offer post-processing ability as well, so if a particular CAM program doesn't offer a needed post-processor, it still may be possible to use that program.

For many hobbyists, this may not be a big issue, because they are creating their own machines and they may work just fine with the basic code produced by a given CAM program. It is good to know what the term means when shopping for

CAM software, though. Remember that the code produced by a CAM machine may not automatically work with a given CNC machine without some modification. G-code is meant to be standard, but isn't quite so.

CNC Controller Programs

The controller program reads G-code instructions and turns them into control signals that are fed to a motor controller box via the printer port or some other output device. A couple of popular examples are the Mach series of controller software (Mach 2 or Mach 3) by Artsoft, or TurboCNC by DAK Engineering which run on Windows XP and DOS respectively. Another one that operates on the Linux platform is EMC (Enhanced Machine Controller). There are many other controller programs available to the hobbyist at a variety of prices, and many of them are quite capable and should perform well once they are properly set up.

For hobbyists, the major advantage of these programs is that they use equipment that many people already have, namely a PC with a printer port. The ones listed are also relatively inexpensive; Mach has a free demo mode, TurboCNC is distributed as shareware, and EMC is freely distributed. Professional CNC machines (big production oriented machines) have their own versions of controller software which comes with the machine.

Constant Velocity Contouring

A phrase you will probably encounter when looking at controller programs is *constant velocity contouring*. This refers to a feature in some CNC controller programs to constantly move and avoid coming to complete stops between lines of G-code during a run. This feature can save machining time by avoiding fully ramping up and down the motor speed between cutting moves. If a toolpath is composed of a lot of short cutting moves, the time savings using constant velocity contouring can be substantial. A drawback of this process is that it is possible to introduce inaccuracies in the tool path if the algorithm is not well implemented.

Toolpath Verification Programs

A toolpath verification program simulates the toolpath the cutting tool takes before running the program on a CNC machine. This allows the operator of a CNC machine to verify that his or her G-code doesn't do anything disastrous like move out of the machine's cutting area or push the tool down into the bed of the machine. It also can indicate if the toolpath is wrong in some other less dramatic way, such as moving in an inefficient way, or cutting a different shape than what was intended.

This is important, because it is as easy to instruct a machine to do something that will damage the machine itself, such as 'crashing' the spindle (slamming the spindle into the mill bed or the stock material), as it is to tell it to do something useful. With an expensive machine, this can result in thousands of dollars of repairs in no time at all. In a hobbyist setting, things will probably not be as dramatic or expensive, but checking new G-code can still save time and some money. Some controller programs (like Mach 3) have basic tool path visualization built in to them, as do some CAD/CAM packages. There are stand alone programs to perform this function as well.

A Word of Advice Regarding Software

Please note that there are many CAD, CAM, and controller programs available in a variety of price ranges. For those new to CNC, I would strongly consider looking for cheap or free alternatives until you figure out what you really need. It is possible (easy actually) to spend hundreds and hundreds of dollars on software that you will rarely use or will only need a small portion of that program's capability. There are some expensive programs that are difficult to use and don't perform all that well. Expensive does not necessarily mean good, and inexpensive does not necessarily mean bad.

Examples of free or inexpensive, but very useful programs available are ACE Converter (an extremely basic CAM program) and the demo version of the Mach 2 and Mach 3 programs (a controller program). Older versions of CAD programs can be purchased for next to nothing, and may serve perfectly well, at least until you get a feel for what you really need. It is wise to begin simply and cheaply unless you really know what you need, so you won't end up with an empty bank account and buyer's remorse.

A Walkthrough of Some Programs

Given the large number of available CAD, CAM, and combined CAD/CAM, packages, and given their complexity, it is not feasible to give coverage of more than a small fraction of what is available. To give you a feel for these programs and how they are used together, a brief illustrated walkthrough is provided for both 2½ D and 3 D machining. The programs I discuss are Rhinoceros (a 3 D modeling program), LazyCAM, MeshCAM, and ACE Converter (three different CAM programs). They vary in price from free to far-from-cheap. Rhinoceros is designed for high precision 3 D modeling but is also capable of creating drawings suitable for 2½ D machining, and can provide pretty much any output format that might be required. LazyCAM is a 2½ D CAM program, as is ACE Converter, and MeshCAM is a 3 D CAM program.

There is nothing magical about this collection of programs. There are dozens of available programs in each category, and as long as the file types are compatible, you should for the most part, be able to 'mix and match'. A high end program such as Rhinoceros, for example, may be replaced by a less expensive program, so long as it can produce an output file that a CAM program can read. Consulting a CNC related list or forum online may help determine if certain combinations will not work, and can provide solutions if problems arise.

The Workflow
Below is a flow chart representation of the workflow for creating G-code through CAD/CAM. Toolpath verification is shown as a separate step, although this may be done as part of a CAM software package, as part of the controller software, or in a standalone program.

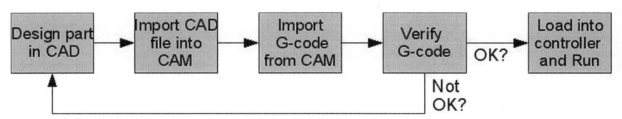

Figure 48: *The part is first drawn in CAD and imported into a CAM program for conversion to G-code. The G-code is usually verified after being created in the CAM program, and if necessary, corrections are made either by hand editing or redesigning the part and regenerating code in CAD and CAM respectively.*

A 3 D CAD/CAM Walkthrough
Step 1 :
Create a Model and Export it
The first step is to create a model in a CAD program and then export it in a format that the CAM program can use. Modeling in 3 D is a bit tricky, and requires some

visualization skills. In general, most programs allow the creation of primitive shapes which may be modified, combined and subtracted to form more complex shapes. For instance, the wheel shown in Figure 49 is a combination of a torus (doughnut shape) to make the outer rim, cylindrical shapes for the hub and spokes, and some small spheres to imitate bolts.

The basics of how to use any particular modeling program may take up a whole book by itself, so if you are interested in learning a specific program, you will need to look for manuals and "how to" material specific for that product.

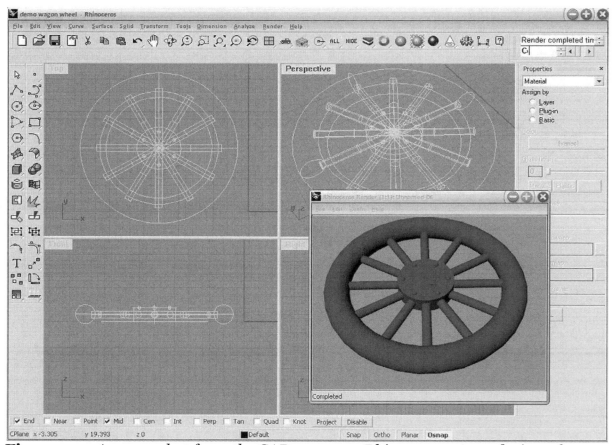

Figure 49: *A screenshot from the CAD program Rhinoceros. A rendering of a wheel is shown in the bottom right corner. The model will be exported from Rhinoceros in the STL file format, which is commonly used in 3 D CAM programs. Although Rhinoceros is intended for 3 D modeling, it may also be used to create precise line drawings which can be exported as a DXF file to use with 2½ D CAM programs.*

Modeling familiar but relatively simple objects is a standard approach to learning how to use modeling software. Many tutorials involve tasks like "create a model of a..." flashlight, car or whatever. These tutorials may be available from the software manufacturer or free on the Internet from the users of a particular modeling program.

Once a part has been modeled to satisfaction, it is exported in a format that the CAM program can import. In this case STL is appropriate, since the 3 D CAM program used here (MeshCAM) has this as its import format. The STL file format is importable into many other CAM programs with 3 D machining capability.

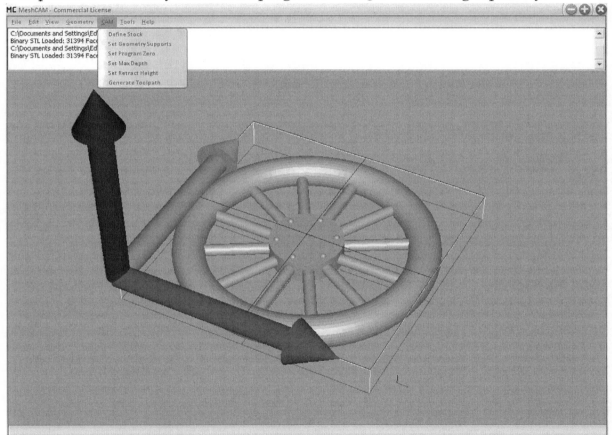

Figure 50: *After importation into MeshCAM, a view of the part is shown. The wire frame box around the part shows the dimensions of the block of stock material the from which the part will be cut. Dropdown menus on the menu bar at the top of the screen allow a numbers of setup parameters to be entered, including the dimensions of the starting material.*

Step 2:
Import into CAM and Generate a Toolpath

The second step is to import the model into the CAM program, set some parameters, and generate a toolpath. MeshCAM provides a 3 D graphical display of the imported CAD model, shows a wire frame representation of the stock material from which it is to be cut, and indicates the X, Y, and Z axes. Among the things that can be set up are how the part is positioned; it may be rotated, shifted, or centered in the stock material. It may also be scaled, so if the operator wants to change the size, this may be done without going back to the CAD program to redo everything.

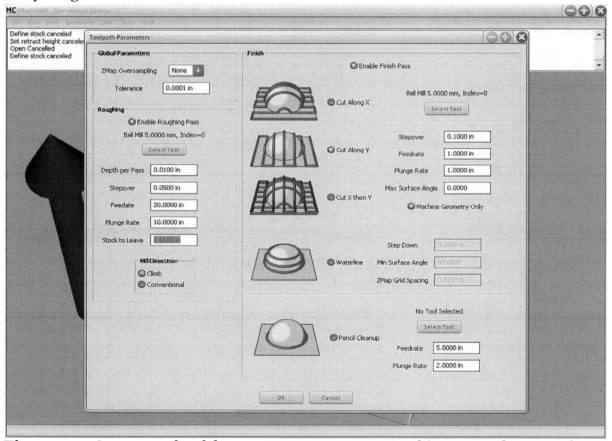

Figure 51: *One example of the many setup screens. On this screen, the type of machining path may be specified, as well as feed rates, whether or not a roughing pass will be made, and how widely spaced the cutting passes are.*

Other details about the run that may be set include the maximum cut depth, as well as the type of cutting tool used. Cutting tools (router bits, end mills, and so forth) that will be used with the machine must be specified; the dimension of each tool, such as the diameter, the radius (if it has a round profile like a ball mill), and the flute length may be entered. Once a particular tool is entered and saved, it may be called up and used for future parts without entering its details again.

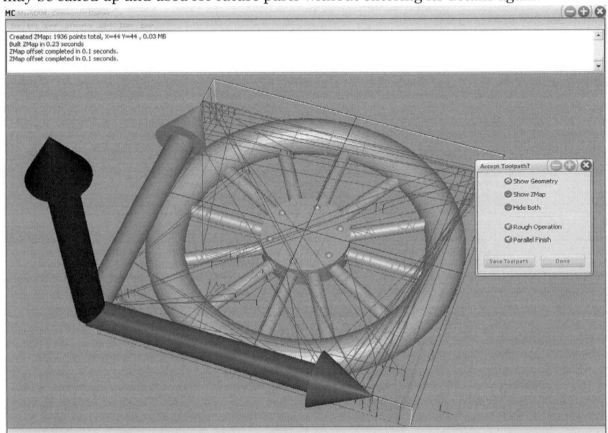

Figure 52: *MeshCAM provides an overlay of the toolpath showing both the cutting moves and rapids (quick repositioning moves) that it has calculated. If the toolpath isn't right at first, parameters can be changed and the toolpath may be regenerated repeatedly until a good result is achieved.*

Prior to generating G-code, toolpath details are specified, including the depth of each cutting pass or the use of climb milling versus conventional milling (defined in Appendix F). The specific type of toolpath to be generated may be set as well.

Different toolpath algorithms have been created to address specific machining situations; some types perform much better for a particular situation. For instance, waterline machining may be suitable for parts with a lot of vertical or nearly vertical surfaces, but doesn't work as well for parts with shallow slopes. More sophisticated and expensive CAM packages provide a greater variety of toolpath types. Additional types become more important in commercial production where higher efficiency and high quality can significantly improve the bottom line.

Once the process of setting parameters is finished, a toolpath can be calculated. The toolpath it creates is displayed for inspection by the user (see Figure 52). Once an acceptable toolpath has been created, the G-code generated by MeshCAM is saved in a file for import into a toolpath verification program or controller program. If some undesirable trait appears during verification of the G-code, then the operator may have to go back to an earlier step to make modifications.

Step 3:
Verify the Toolpath and Refine it if Necessary
Toolpath verification of the G-code may be performed using a standalone program, or this function may be included in a more full featured CAM or CAD CAM programs. These programs usually provide a 3 D visualization of the stock material and the cutting tool, and show how the tool traverses through the stock when cutting. They also can calculate how long the entire machining a run will take. Other features, such as collision detection and a G-code editor may be available as well.

Although it is primarily a controller program, Mach provides basic visualization of the toolpath once the G-code is loaded, which may be inspected prior to a run. It can also step through the G-code and calculate the run time.

Regardless of what program is used for verification, the basic idea is to inspect the toolpath to see if it does anything funny, such as making a move outside of the cut area of the machine, making cuts other than those intended, or following an inefficient (unnecessarily slow) path. If problems appear, the G-code can modified by hand or reworked using the CAD and CAM programs until everything is fixed, *prior* to actually trying it out on the machine.

A 2½ D Walkthrough

The process with the 2½ D CAM program is similar to that with the 3 D program, with a few exceptions. As mentioned before, many 2½ D CAM programs import DXF files, which represent a blueprint style (flat, two-dimensional) line drawing.

Step 1:
Create a Drawing in a CAD Program

We will again begin with our CAD program (Rhino), but we will just use the basic curve drawing tools available to make a line drawing of the desired part to scale. The part being created is the face of a speaker cabinet composed of two holes (for the woofer and tweeter), a decorative fish design carved into the face, and the exterior rectangle of the face. The drawing is shown in Figure 53.

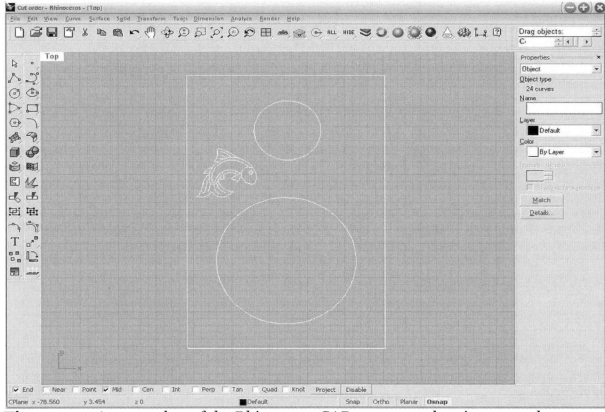

Figure 53: *A screenshot of the Rhinoceros CAD program showing a speaker face. Two holes will be cut all the way through, a decorative fish design will be carved at a shallow depth in the face with a V-shaped bit, and finally the outer rectangle will be cut through the material. The drawing is saved as a DXF file.*

Two programs for CAM will be covered this time, namely LazyCAM and ACE Converter. LazyCAM is produced by the same company that creates the Mach series of controller programs, and ACE Converter is a simple, freeware CAM program for converting DXF files into G-code, and is distributed by DAK Engineering. When using Ace Converter, different parts of the drawing are placed in different *layers* to allow different parts of the drawing to be cut to different depths and in a specific order. Layers is a feature in many drawing programs that allows the user to group different parts of a drawing together, allowing them to be distinguished from each other and manipulated. This is not required for use of LazyCAM, however, which allows the user to set depth information for different parts of the drawing after the file has been imported into the program.

Step 2:
Import into CAM Program and Set Parameters

Importing into LazyCAM

In the case of LazyCAM, the imported DXF file shows up as a flat line drawing which should look like the drawing produced in the CAD program (refer to Figure 54). Features of the drawing are selected, and both the cut depth and order in which that feature will be cut may be set. More complex parts will have more features, and each will require a cut depth and cut order.

In this case, all of the lines that make up the engraving of the fish pattern are assigned to be cut to a very shallow depth, since it is meant to be a carving on the face, and not meant to be cut through the material. To cut out the panel, the outer rectangular will be assigned to cut to the depth of the stock material (actually, slightly deeper so it will cut all the way through), and so will the holes in the interior. As with the 3 D CAM program, it may be necessary to set a variety of other parameters (tool dimensions and so forth). Once this is done, the toolpath can be calculated and G-code can be produced.

Figure 54 (following page): *A screenshot of the program LazyCAM, showing the DXF file generated from the CAD program. The individual features of the drawing (referred to as 'chains' in LazyCAM) can be selected and assigned a cut depth and cut order. The holes and the outer rectangle will be cut all the way through the stock (about ¾"), but the engraved fish pattern will be set to cut to a shallow depth.*

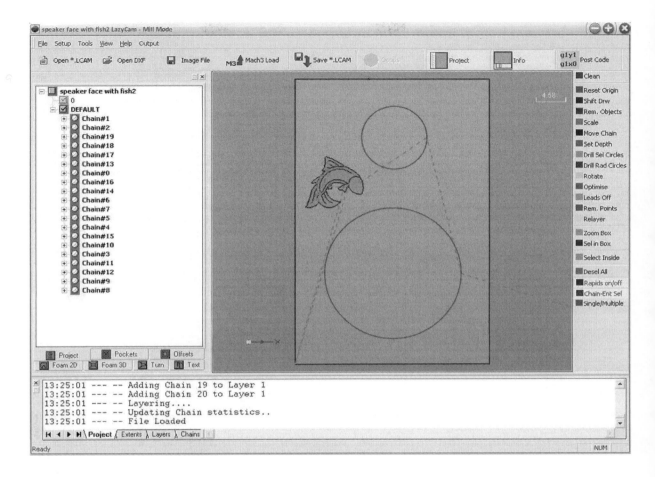

Ace Converter

Please note that these screen captures are from Version 3 of Ace Converter. The Version 4 beta has just been released, and appears to have a revamped interface and an expanded feature set. Things may look a bit different depending on which version you use.

Ace Converter has been around for a while and is very basic program. This is not to say that Ace Converter isn't useful, however, it is and the price (free) can't be beat. Ace will take a DXF file and will kick out G-code based on it. As mentioned before, operation of the program requires that different 'layers' be specified when drawing the part in the CAD program. Cut order and cut depth are assigned to the different layers in the program.

The interface for Ace Converter is very simple, which makes it easy to learn, but it requires that the user remembers what part of the drawing is assigned to each layer. Unlike LazyCAM, this version of Ace Converter does not show what the loaded DXF file looks like, so you are flying blind to some extent until you load the G-code in a toolpath verification program. A newer beta version of Ace Converter has been released, which appears to have a new interface and some capacity to visualize the part that is being cut, so it is possible that the screens (and capabilities) of the version that you use will be somewhat different than described here.

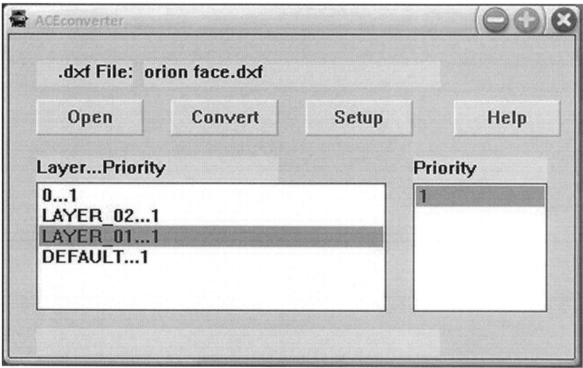

Figure 55: *The main screen for Ace Converter...yes, that's all there is to it. This screen lists the individual layers in the box on the left. Double clicking on a layer will bring up the screen as shown in Figure 56; a depth of cut is assigned to all of the shapes from the drawing that are on that layer. The simplicity of Ace Converter requires that some additional work be done in the CAD program to place different features of a drawing onto different layers. It is a useful program and a good place for a new user to start, however. G-code is produced by pressing the 'Convert' button.*

The conversion process does not always work as expected, and some odd results may appear (such as having an arc drawn improperly). If this happens, it may be necessary to re-draw the part in the CAD program and re-import it into the CAM program for conversion. Sometimes simple things may be corrected by hand editing the G-code using the text editor.

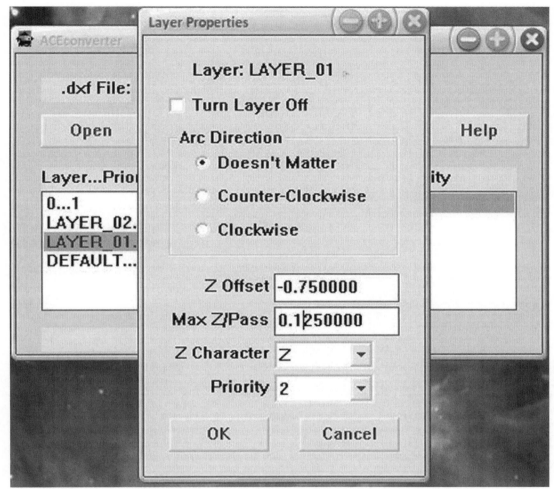

Figure 56: *Layers may be selected and assigned a cut depth and order (priority). In this case, the cut depth is set to .75 inches, and the maximum depth of cut per pass is .125 inches. Six passes are necessary to cut to the desired depth, with each pass deeper than the previous. This process is repeated for each layer specified in the DXF file.*

Step 3:
Verify the Toolpath and Refine it if Necessary

This step is pretty much the same as step 3 in the 3 D CAM walkthrough. First, import the G-code produced by the CAM program into some kind of program with a toolpath verification or visualization function. Then inspect it for wrong or damaging moves, see if you are cutting the part you think you are, and check for an inefficient (i.e. poorly laid out) toolpath that could to be improved to shorten the run time of the program.

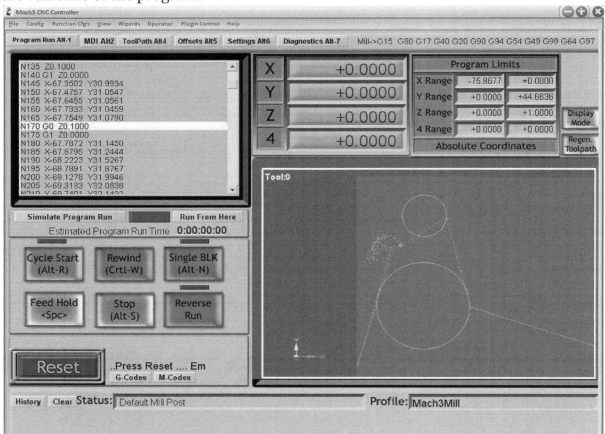

Figure 57: *A screen capture of the G-code exported from LazyCAM into Mach 3. The toolpath is visualized in the lower right corner, and is how Mach 3 interprets and will cut the G-code file (which is shown in the upper left).*

This leads us to the final step in the journey, which is to run polished and quality checked G-code in a controller program. The program described here is the previously mentioned Mach controller program. The first time the controller program is used, there is a lot of set-up to get the controller program to communicate properly with the controller box and to configure it properly for the machine in question. This is described in the following section.

Setting Up the Controller Program

As with CAD and CAM programs, it would be exceedingly difficult to review all of the available controller programs, given the large number of choices. To get a feel for what these programs do and what features they offer, a walkthrough of Mach 3 is provided. As with the previous programs, screen captures will be shown and aspects of operating Mach 3 will be described.

The Mach series of controller programs (Mach 2, Mach 3, and so forth) operate under Windows XP, and are widely used by CNC hobbyists. These programs have free demo modes that allow programs of up to several hundred lines of G-code to be run. As of this writing, different versions have different maximums. Regardless, this is enough for substantial experimentation.

A certain amount of configuration is required for proper communication between the controller software and the motor drives. Usually, once the controller software has been successfully configured to work with a particular CNC machine, little else needs to be done for future runs other than to load G-code and run it, unless some feature is changed on the machine in the future.

It is important to properly designate the parallel port pins that are used to control the motor drives and other devices. If a particular set of pins have been wired to control the X-axis motor drive, for instance, this must be configured in software to match how they are physically wired. Recall that some pins are only for output, and others are only for input. Input pins may be used for tasks such as stopping the machine when a limit switch is triggered. The ones used to control the motor drives are output pins. Each motor drive requires two pins to operate; one for the step signal and another for the direction signal. A connection to the ground on the parallel port is required as well, but this can be shared by several drives.

Motor Set Up

The controller software must be told how many steps there are to a unit (inches or millimeters are common). For instance, if a motor / motor drive / lead screw combination requires 2000 steps to create a movement of one inch, then this value must be specified for the controller software to work properly. This may be done individually for each motor, allowing great flexibility in set-up. Different motors (with different steps per revolution) and lead screws (with different leads) may be used together on the same machine.

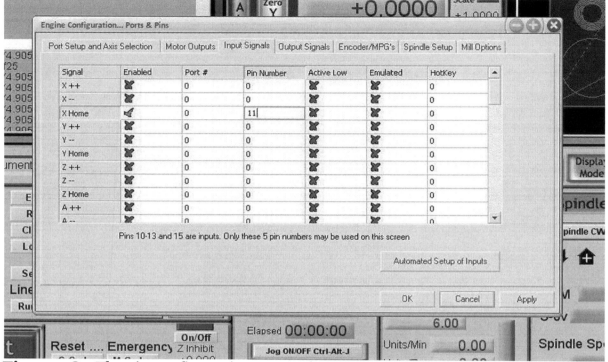

Figure 58: *The pin configuration screen from Mach 3. It is here that specific printer port pins are designated to provide step and direction signals to the motor drives. In this case a total of six pins will be used to control three motor drives; one pin for each step and direction signal for a given motor. These pin designations must correspond to the actual wiring from the printer port to the motor drives to work properly.*

Acceleration Profile

An acceleration profile should be set for each motor on the machine. This allows the operator to configure how quickly a motor accelerates and decelerates, as well as its top speed. This can be very useful in fine tuning the performance of the CNC machine based on its specific motors, motor drives, and the conditions under which that machine operates. Different setups and different types of machining operations will probably require some tweaking to operate best.

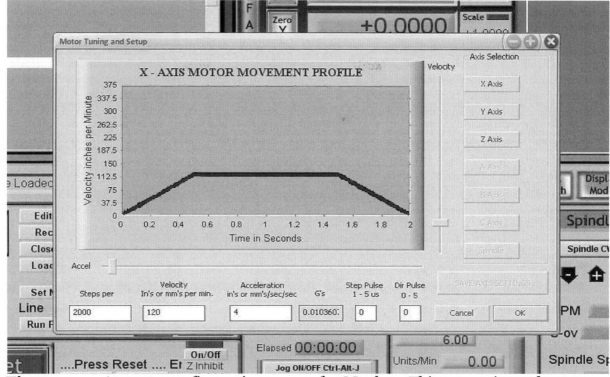

Figure 59: *A motor configuration screen for Mach 3. This screen is used to set the number of steps per unit of movement (in this example 2000 steps result in one inch of movement). Acceleration and deceleration rates and the top speed may also be set. A different profile may be set for each motor that is used.*

Duration of Step and Direction Pulses

Another variable that may be configured is the duration of the step and direction pulses. Different motor drives have different requirements for the duration of an electrical pulse for it to recognize the pulse and act on it. This duration may be set in software to help insure that the controller program will work properly with different motor drives. Motor drive manufacturers will usually provide information about minimum and maximum duration of a signal.

Additional Setup

There are many additional things to configure in controller software such as which keys on the keyboard are used for jogging the motors into position, how G-code is interpreted, and how a fourth or fifth axis is used. Some of these parameters may be irrelevant to the use of a certain machine if that machine doesn't have that particular feature. Documentation should be available from the source of the program to explain how to properly configure all available features. There also may be tutorials or FAQs available for the controller software you are using on how to do this, or an online forum for a specific piece of software.

After the initial setup is finished or at the point when you think it is finished, the controller box can be connected and the motors tested. It is a good idea to first test for proper function of the motors and electronics prior to connecting them to the machine, so that there won't be any major surprises if things aren't configured properly.

Running a Controller Program

After the set up is completed, operation of the controller program involves importing a G-code file into the program, setting the start point for the machine using home switches or by jogging the machine into position, then allowing the controller software to 'do its thing'. Once the controller has started running, the software interprets the instructions in the G-code and sends streams of step and direction signals to the motor drives, resulting in the desired motor movement.

During operation, the position of the cutting tool is displayed on the main screen, along with the currently executing lines of G-code. The main screen has start, stop, hold, and reset buttons, among others to control the program run. Refer to Figure 60 for a screen capture of the Mach 3 run screen. It contains a lot of the information that a typical CNC display shows during a run.

Final Comments

The chain of software leading from your design to a part cut on a CNC machine is a bit long. A good way to start producing G-code would be to draw a simple shape in a CAD program, import the file into a CAM program such as ACE Converter, to produce some G-code. The code can then be imported into the toolpath verification program to see what the toolpath looks like.

It is also a good idea to take some G-code that you've produced using the method just described, or that you've found somewhere else and try editing it in a simple text editor. Feed rates, moves, and other things can be changed, and then the code can be examined using toolpath verification to see if things changed as expected. The next chapter gives an introduction to the basics of G-code.

Ultimately this CAD, CAM, and G-code verification process is a typical way to create something using CNC. The editing step is common too; sometimes it is necessary or easier to make a few simple changes by hand to get things to work right. After the first few tries it should become fairly obvious how things come together.

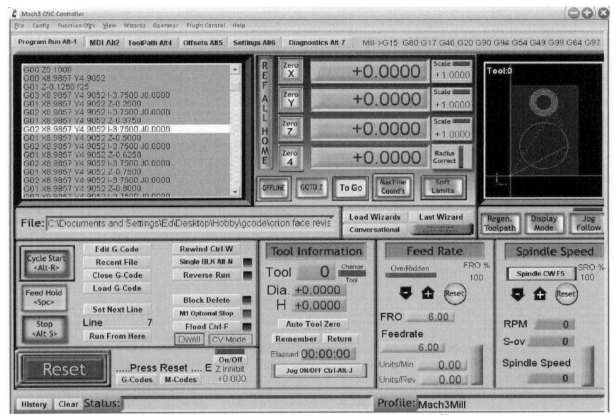

Figure 60: *The Mach 3 'Program Run' screen. CNC controller programs commonly display the machine coordinates and the current lines of G-code being executed, as well as the feed rate and what the current tool is (if it has an automatic tool changer). A keyboard is used to execute basic control the CNC machine prior to starting, such as positioning the cutting tool machine (referred to as 'jogging' the machine), and is also used for entering values for X, Y, and Z-axis position, feed rate, and so forth.*

15

G-Code

```
G00 Z0.1000
G00 X2.2951 Y1.7060
G01 Z-0.1200
G03 X2.2951 Y1.7060 I-0.2423
J0.0000
G01 X2.2951 Y1.7060 Z-0.2400
G02 X2.2951 Y1.7060 I-0.2423
J0.0000
G01 X2.2951 Y1.7060 Z-0.3600
G02 X2.2951 Y1.7060 I-0.2423
J0.0000
G01 X2.2951 Y1.7060 Z-0.4800
G02 X2.2951 Y1.7060 I-0.2423
J0.0000
G01 X2.2951 Y1.7060 Z-0.5500
G02 X2.2951 Y1.7060 I-0.2423
J0.0000
G00 Z0.1000
G00 X5.0965 Y2.5447
G01 Z-0.1200
G01 X5.0965 Y2.5447 Z-0.1200
G03 X5.0965 Y2.5447 I-1.6250
J0.0000
G01 X5.0965 Y2.5447 Z-0.2400
G02 X5.0965 Y2.5447 I-1.6250
J0.0000
G01 X5.0965 Y2.5447 Z-0.3600
G02 X5.0965 Y2.5447 I-1.6250
J0.0000
```

The Language of Machine Tools

Someone who wants to run a CNC machine should know something about G-code. 'G-code' actually refers to a portion of the entire programming language used for numerical control (there are 'F', 'S', and other codes as well), but this term is commonly used to refer to the whole numerical control programming language. It is used to specify the locations, movements, and actions that the operator wants the machine to perform.

G-code is a large subject, and this is enough of an introduction to get you started, but hopefully not so large as to get you confused. There are many tutorials and useful reference sites on the Internet, as well as several detailed books on the subject; I would suggest looking to these resources when you advance to the point where you need more detailed instruction.

As you progress you will probably end up using CAM programs to generate most of your G-code automatically. That being said, the goal here is not so much as to learn how to write complex G-code from scratch, but rather to learn how to interpret and edit it. The basic skills you learn early on will allow you to inspect and modify the G-code that is created by a CAM program.

The basic process for hand coding is pretty straight forward:

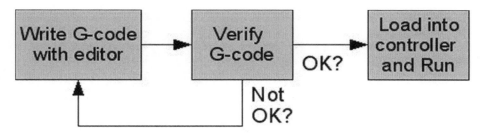

Figure 61: *G-code is written, verified, and re-edited if necessary before being loaded into the controller software. You can also take G-code output from a CAM program and modify it by hand.*

Basic Commands

As described previously, the instructions that are loaded in a CNC controller program are written in G-code. G-code is relatively easy to read and edit. A few of the most fundamental G-code instructions are listed below. Please note that the circles in the commands are zeros and not the letter "O".

Code	Description	Comments
G00	This positioning move made at a relatively quick speed (i.e. the 'rapid' rate).	Used to reposition the cutting head quickly.
G01	A linear interpolation made at the current feed rate.	A straight line move made starting from the current spindle position.
G02	A clockwise circular interpolation.	A move through an arc or full circle with a specified radius in a clockwise path.
G03	A counter-clockwise circular interpolation.	Same as G02, except in a counter-clockwise path.
F	The F word specifies the feed rate; the rate that a cut is made.	All cutting moves (such as G01 and G02/G03) will cut at the last feed rate set.
M0	Instructs machine controller to stop the program.	One of many 'Miscellaneous' codes used to perform many different functions.
S	Sets the spindle speed.	For those who have machines with computer controlled spindle speed.

The term *interpolation* implies that the controller program determines the intermediate points for a move which start at the current position and ends at the position specified in the line of code. Read the terms 'linear interpolation' and 'circular interpolation' as 'linear move' or 'circular move' with the understanding that you specify the start and end points, and the controller program figures out the stuff in the middle.

Modal Commands

Some commands in G-code remain in effect until you specifically reset them, that is to say, they are *modal*. For instance, setting the feed rate to 15 in a given line of code using the F command means that this feed rate will be maintained for future moves until it is reset with another F command to a different value. Successive G01 commands (and other move commands such as G02 and G03) will move at a feed rate of 15 until things are changed. You can think of modal commands as operating like the Caps Lock key on a keyboard. Once you set it, it will stay on until you deliberately reset it.

An exception to this is the G00 command. It is set to move at whatever the 'rapid' move rate is (the speed set for quickly repositioning the spindle). This rapid rate is configured in the controller software and is often set close to or at the machine's maximum speed.

A Simple G-code Program: Cutting a Square

G-code may be written in a simple text editor (such as Word Pad or Note Pad in Windows). Using a word processor may cause problems because it may add extra characters for formatting and such to the file. Any program that can save as a simple text (.txt) file should work, however.

A very short G-code program to cut a square path is shown below with lines of code listed on the left, and comments on the right. We will presume here that the units in this program are inches and inches per minute (for the feed rate), but in reality the units are whatever they are set up to be in the controller program. Sometimes G-code is written with line numbers, but these examples will exclude them.

Code	Comments
G00 Z0.100	Move Z-axis to +.1 inch (so cutter isn't in the material).
G00 X0.000 Y0.000	Move to X = 0 and Y =0 at the rapid rate of travel.
G01 Z-0.100 F20	Move Z to -.1 inch (move into material at feed rate 20).
G01 X2.000 Y0.000 Z-0.1000	Move to X = 2 inches (cut a line from 0,0 to 2,0).
G01 X2.000 Y2.000 Z-0.1000	Move to Y = 2 inches (cut a line from 2,0 to 2,2).
G01 X0.000 Y2.000 Z-0.1000	Move to X = 2 inches (cut a line from 2,2 to 0,2).
G01 X0.000 Y0.000 Z-0.1000	Move to Y = 2 inches (cut a line from 0,2 to 0,0).
G00 Z0.1000	Move the cutter to a safe height above the material.

The program begins by moving the cutter (end mill, router bit or whatever) to a position of Z = .1, a position .1 inches above the surface of the material being cut. The spindle is then repositioned to a start position (0, 0 in this case) and then is moved down into the material at a feed rate of 20 inches per minute.

From here the next four lines tell the cutter to cut a square, moving from the coordinates 0,0 to 2,0 to 2,2 to 0,2, and then back to 0,0. The final line then lifts the cutter out of the material to a height of .1 inch above the material. Please note that each successive line will move the cutter from the final position reached in the previous line.

To reiterate, the difference between G00 and G01 is that G00 is a quick positioning move, and as described before, will travel at the top speed specified in the control program (known as the 'rapid' rate) regardless of what the previous feed rate was. A G01 command, however, will move at whatever speed was set at the last F command. So if a G01 command is completed at a feed rate of 20 inches per minute, and then a G00 command is used to reposition the spindle, this repositioning move will be made at a faster rate than the feed rate. If another G01 move is made (or G02 or whatever) it will move at whatever the last feed rate was. In this case, 20 inches per minute.

Deep Cuts Require Several Passes

To make a deeper cut into a piece of material, successive passes are made by repeating what is largely the same block of code. For example to cut this square to a depth of .2 inches, the following code would be used:

Code	Comments
G00 Z0.100	
G00 X0.000 Y0.000	
G01 Z-0.100 F20	
G01 X2.000 Y0.000 Z-0.100	
G01 X2.000 Y2.000 Z-0.100	
G01 X0.000 Y2.000 Z-0.100	
G01 X0.000 Y0.000 Z-0.100	First pass is completed.
G01 X2.000 Y0.000 Z-0.200	Move Z down to -.2 and repeat same pattern.
G01 X2.000 Y2.000 Z-0.200	
G01 X0.000 Y2.000 Z-0.200	
G01 X0.000 Y0.000 Z-0.200	Second pass is completed.
G00 Z0.100	

In general, deep cuts are made with several passes, specifying a lower (more negative) Z height for each successive pass. To cut to a depth of .75 inches, eight passes would be required if the maximum cut depth for a given pass is .1 inch (seven passes at .1 inch depth and a single pass at .05 inch). The maximum depth that you can cut per pass is determined by several things, including the diameter of the bit used, the nature of the material (hard vs. soft), how powerful the spindle being used is, and the feed rate. Chapter 16 gives more detail on which factors influence the maximum feed rate.

Cutting a Circle

Cutting straight lines is as well and good, but curves are useful too. Circular paths are coded using the G02 and the G03 commands. Sample code for cutting a circle is given below:

Line	Comment
G00 X2.000 Y1.000	Make a rapid move to X = 2 and Y = 1.
G01 Z-0.100	Cut down into material .1 inch.
G03 X2.000 Y1.000 I1.000 J1.000	Make a circle starting and ending at X=2, Y = 1 centered at X=1, and Y=1.
G00 Z0.100	Move Z-axis to a safe height above work piece.

An Explanation of the Code

To start, assume that the cutting tool is at a safe height above the surface of the material to be cut. It moves to the start position at X=2 Y=1, and moves down into the material .1 inch. It then cuts a circle from the starting position and ending at the same place, and is centered at X=1, Y=1. In the line

G03 X2.000 Y1.000 I1.000 J1.000

The X and Y coordinates represent the *ending* position. The starting position is the last position that the tool was at, which is also X=2 Y=1 (which was set up by the first line of code). The I and J values are the center of the arc, which is a full circle in this case. The difference between G02 and G03 is that they specify clockwise and counterclockwise travel, respectively.

G02 and G03 may also be specified with an R parameter (for radius). For example, a line of G-code such as:

G02 X3.0000 Y1.0000 R5.0000

tells the machine to cut from whatever the current position is to X=3 and Y=1 in an arc with a radius of 5. The above line of code would move in an arc and not a complete circle in most cases.

M-Codes

The letter M specifies 'miscellaneous' codes, such as codes to start and stop the spindle, and to end the G-code program. For instance, M05 is the code to turn the spindle off, and M30 indicates to the controller software that the G-code program has ended and should rewind back to the beginning. With professional equipment, M-codes vary somewhat by manufacturer. Due to their relative simplicity, DIY machines may use a very limited number of M-codes.

There are a specific set of G-codes available that allow the user to define a series of offsets (relative home positions). This is convenient when cutting replicate parts on a CNC machine. The G-code that is used for one part may be 'recycled' by simply giving it a different start point elsewhere in the cut area of the machine.

The G-codes for these offsets are listed below:

Code	Description	Comments
G53	Move referenced from machine home	This is a machine coordinate referenced from the machine home, i.e. where X=0, Y=0, Z=0 refers to the machine home.
G54-G59	A move referenced from a user specified coordinate system	A number of coordinate systems may be specified, where G54, G55, G56, G57, G58, & G59 are user defined home (X=0, Y=0, Z=0) positions.

For instance, if the user defines a coordinate system for G57 relative to machine home of X=2, Y=1, Z=0, then all G57 moves would be made relative to this position, so a line of G-code such as:

G57 X0.500 Y1.000 Z0.000

tells the machine to move to (X=0.500, X=1.000, Z=0.000) in the G57 coordinate system, which is X=2.500, Y=2.000, Z=0.000 relative to the machine's home position. G54-G59 may all be defined to provide a number of different zero points relative to machine home, providing flexibility and convenience when cutting the same part repeatedly or creating complex parts that have the same pattern cut in different locations on the work piece.

Tool Offsets

Cutting through a piece of material using our original example G-code program (the one that cuts a square of 2 inches per side) will most likely result in a square that is not 2 inches on each side. This is because as written, this program moves the *center* of the cutting tool in a path that is 2 inches per side. The tool has a width that must be considered. If you were to use a .25 inch diameter bit to cut the square, the square that is cut will only be 1.75 inches on each side because the tool removes half of its diameter on either side of its center when cutting. To produce a square with sides of 2 inches, the path cut must be 2.25 inches per side, to account for the .25 inch width of the cutting tool. CAM programs allow you to specify tool dimensions, and will automatically compensate for the cutter dimensions when generating G-code from the input file. This feature of CAM programs makes this a relatively trivial issue to deal with. However, if you are hand writing G-code, you will probably need to adjust for this manually.

Absolute and Relative Coordinates

The code examples thus far have been in an absolute frame of reference, meaning that the positions specified in the code are in relation to a fixed starting point (home, or zero point). It is also possible to operate in a *relative* (also known as *incremental*) coordinate mode, where moves are made as if the current position is the zero or home position. For instance, taking the code from the first example:

G01 X2.000 Y2.000 Z-0.100

In absolute coordinate mode, the final position will be X=2, Y=2 from the home position (0, 0), regardless of the starting position. In a relative coordinate mode, this would move a distance of +2 in the X direction and +2 in the Y direction from the current position.

If you are confused about using relative coordinates, then it is probably best to stick with absolute coordinates, at least initially, because relative coordinates tend to be more confusing for most people. A little more will be said about relative and absolute coordinates in the next section.

A Final Word

There are many tutorials available on the Internet, and several hundred page books on G-code are available, but this intro should provide a push start. It is possible to write code to create very sophisticated parts. In reality, however, the goal isn't to learn how to code complex parts by hand, because this is what CAM programs are for, and even hand-calculating how to cut some relatively simple arcs can be a bit tricky and prone to error. Knowing how to read and write some G-code, however, gives the operator a chance of troubleshooting problems and tweaking things when the need arises. This is a case where a little knowledge will go a long way.

16

Using Your CNC Machine

Using Your Machine

These subjects are somewhat random and are really just a collection of tips, tricks, and mini-tutorials. Consume as needed. Included are discussions of:

- Machine coordinates
- Cutting tools
- Feed rates and forces
- Speed requirements
- Setting the tool height
- Cut order
- Precision, accuracy, and reproducibility

Machine Coordinates

CNC machines typically specify three axes, creatively labeled X, Y, and Z. The X and Y axes refer to the plane that a work piece is being held to, and is usually the plane parallel to the ground. The Z-axis is usually the vertical axis; the one that moves perpendicularly to the ground (and the X and Y plane). You can refer to the axes any way you want, but this is typical. The X-axis is usually the axis with the longest travel. Again, this is an arbitrary but typical designation.

In the case of a mill, where the work piece is moved relative to the cutting tool (where the spindle is stationary with the exception of up and down movement), a move in the positive direction along either the X or Y axis means that the bed holding the work piece moves in the negative direction. In machines where the work piece is stationary and the cutting tool moves relative to it (as in a CNC router with a gantry), a move in the plus direction means that the cutting tool moves in the plus direction.

It is common to make the lower left hand corner of the X and Y axes the home (X=0 and Y=0) position. This makes the entire cut area positive (the X and Y values are always positive). It isn't essential to do it this way, but it is easier for a lot of people. For the Z-axis, a move in the positive direction refers to moving upwards, and negative values indicate a move downwards (towards the material being cut).

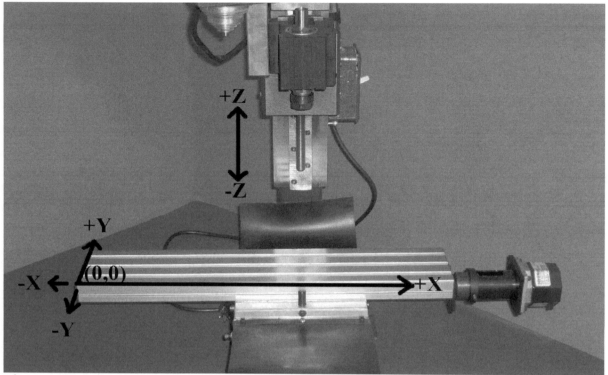

Figure 62: *A common coordinate convention; (X=0 Y=0) represents the lower left hand corner, and the coordinates of the part being cut are more positive to the right (in the X direction) and toward the back of the mill (Y direction). Looking down at the mill table from above, it is basically the same as the upper right quadrant of the Cartesian coordinate system. Positive Z movement is up, and negative Z movement is down (toward the part). Note that if a cut is being made in the positive X direction, the mill bed travels in the opposite direction (towards the left), because movement is determined relative to the part being cut. This is not the case with a gantry system like the one discussed in Chapter 11, where the spindle moves relative to the stock material which is fastened to the stationary table.*

Relative and Absolute Coordinates

As discussed in the previous chapter, a CNC machine operator should understand the difference between relative and absolute coordinates. Absolute coordinates are always made relative to a fixed zero position, known as "home" or (X=0, Y=0, Z=0). This spot may be arbitrarily designated by the user in the controller software, or it may be done using home switches. Home switches are built into

CNC machines to provide an absolute reference position which may automatically be located by the controller software. Prior to the first run of the day (or whenever desired) the CNC controller software can be set to seek the home position using the home switches, and should be able to reliably reestablish the same spot repeatedly.

Relative coordinates indicate that a move is made relative to the current machine position, and not from the absolute zero position. For example, if you tell the machine to move one inch in the positive X direction, it will move one inch from its current X position and not to X = 1. Another way to look at this is to view the current machine position as home, and the next move is made relative to this position.

Cutting Tools

Most CNC machines use a cutting tool of some sort, ranging from the mundane (a drill bit) to the exotic (a laser). For most hobbyist CNC machines, however, a cutting tool will usually be a drill bit, router bit, or end mill.

Most cutting tools of this type are manufactured from either High Speed Steel (HSS) or carbide. Carbide refers to tungsten carbide, and high speed steel refers to a specific alloy of steel formulated to withstand high temperatures. Carbide cutting tools are harder than HSS tools and can withstand higher temperatures. They are much more expensive than HSS, however. Often, carbide cutting surfaces are attached to steel bodies (referred to as 'carbide tipped') to provide some of the advantages of carbide at lower cost than solid carbide tools.

It's smart to purchase relatively inexpensive cutters until you get a feel for what cutting tools you use frequently. Individuals new to CNC machining and machining in general may discover that their interests and needs become better defined, or change fairly quickly after they start using their machine. Also, a new user may be a bit pardon the pun more prone to damaging tools until they have some more practice using their machine. Given the cost of some end mills, it is wise to start out inexpensively until you get a better feel for what you really will be using on a regular basis. At that point it will make more sense to invest in better, more expensive cutting tools.

End Mills and Other Cutters

The term end mill refers to a wide array of cutting tools used in industrial applications, as well as with hobby level milling equipment. End mills, unlike a drill bit for instance, are often used to cut laterally (side to side) through material instead of just vertically, and can produce a variety of cut profiles, such as a square bottom or dovetail. Specialized end mills are available for a variety of purposes. A ball end mill (a cutter with a ball shaped cutter profile), for instance, might be used in a situation where a smooth, curved part is being cut.

Figure 63: *A collection of end mills and other cutters in holders for use with a commercial grade mill. Cutters vary in diameter, cutting profile, number of flutes, and purpose, such as for roughing out material or producing high quality final cut. Quality cutters in good condition can cut significantly faster and cleaner than lesser ones, and may be worth investing in if they are going to be used heavily.*

Materials, Forces, and Feed Rates

It is pretty obvious that steel is more difficult to cut than aluminum, and aluminum is much more difficult to cut than a piece of particle board or foam. In general, the harder the material you are trying to cut is, the more resistance that material will put up to a cutting device moving through it. This resistance is transferred back to the structure of the CNC machine. This is why professional milling machines are big brutes, with massive amounts of metal in their structures to make them rigid. They must be able to withstand the forces generated by rapidly cutting through large volumes of hard material without flexure in either the frame or the components on the machine. Further, they must be able to generate a lot of force to push their way through these difficult cuts, and their spindle motor can't bog down easily or it will stall.

This is also why something like a foam cutting machine get away with using light duty (but otherwise reasonably precise) components. The cutting procedure (melting foam with a hot wire) requires very little force. This means that very little force is transmitted back to the structure of the CNC machine during operation, so relatively light duty components may be used successfully.

A mini mill may be able to make some of the same parts as a professional machine, but will do so in a fashion that doesn't generate the forces that a professional machine does. It will have to make slower and/or shallower cuts. This means that making a given part out of a relatively hard material (like aluminum) on a light duty machine will take a very long time. It is possible, but slooooow. This is for a number of reasons, including the rigidity of the machine itself, the power of the spindle, and the power of the motors driving the axes.

What Determines Maximum Feed Rate?

The maximum possible feed rate (i.e. movement rate when cutting) on a CNC machine is determined by several things, including the type of material being cut, the depth of the cut, the power of the spindle doing the cutting, the type and quality of the cutter being used, and the maximum possible torque of the motors being used to drive the machine. Harder materials (e.g. aluminum vs. plastic) require slower feed rates for a given depth of cut. Deeper cuts require slower feed rates for a given material. More powerful spindles, such as a 3 HP router instead of a 1¾ HP router on a router table, will allow the operator to cut at a higher feed rate in many situations because more powerful spindles don't stall as easily when making a demanding cut.

The following table summarizes factors that influence the *maximum* possible feed rate when cutting through material, everything else being held equal:

If this factor...	Increases	Decreases
Hardness of material	Feed Rate Decreases	Feed Rate Increases
Depth of cut	Feed Rate Decreases	Feed Rate Increases
Diameter of end mill or bit	Feed Rate Decreases	Feed Rate Increases
Power of spindle used	Feed Rate Increases	Feed Rate Decreases
Sharpness of end mill or bit	Feed Rate Increases	Feed Rate Decreases

The Effect of Bit Profile
The profile of a cutting tool can also have a substantial influence on the maximum feed rate. A "V" shaped bit, for instance, may cut at a faster feed rate than a straight bit when cutting to the same depth, because the V bit may be cutting a smaller cross sectional area than the straight bit. The volume of material that a cutter has to clear in a given amount of time is in large part what influences how fast it can move while cutting. An exception to this rule is that that some specialized bits (e.g. ones with less rugged construction than a regular straight bit) may require extra care; namely slower feed rates during use to avoid damage.

The Need for Speed

Many people who are new to CNC often become obsessed with how fast their prospective CNC machine will go. It would be nice for everyone to be able to have rapids of 500 inches per minute, but for many applications (especially the ones that hobbyist types do) insane speed isn't entirely necessary.

Regardless of how fast your machine can move, you probably will not spend more than a fraction of time having it move at top speed. The limiting factor in many cases is how fast it can move *while cutting something*. Having a cutter go blazing through some hard to cut material may result in a stalled spindle and/or a broken cutting tool. Even if a machine can make a cut at high speed, it may not make a particularly clean cut doing so.

This is not to say that high speeds aren't nice or necessary in some situations, but high performance in this area typically requires an investment in heavier motors, heavier materials for the frame, and a more powerful spindle (depending on the application). The take home message is that you need to consider a machine's intended use and your budget when considering how fast to make it go.

If a machine will spend a lot of time repositioning, i.e. moving from spot to spot while not cutting anything, then an emphasis on speed may be warranted. This is because repositioning moves may be made at top speed, and so a higher top speed will result in more substantial time savings than if a machine spends most of its time slowly cutting some material. It is also true if a machine will be doing a task in a production fashion (i.e. you are using it to produce parts for sale) then how fast you produce whatever it is you are making can really matter, because in this situation the bottom line may be impacted substantially.

Getting Ready to Cut: Tips, Tricks, Considerations

Cutting parts on a CNC machine requires a certain amount of prep work and strategy, even with relatively simple parts. As you use a CNC machine, some of these things will become second nature, but it doesn't hurt to have an explanation ahead of time.

Finding Zero

Prior to actually cutting a part, it is necessary to determine the proper zero point for the X, Y, and Z axes. Finding zero for the X and Y axes is not a big deal. It may be set arbitrarily in the software, or may be located automatically using home switches. Most of the effort goes into setting the zero point for the Z-axis, which may vary from run to run depending on the stock material being cut.

In 2½ D and 3 D machining, one practice is to set the zero point for the Z-axis as the top surface of the material to be cut. To further explain, the zero for the Z-axis is the point where the bottom of the cutting tool just barely touches the top of the material to be cut. If multiple cutting tools are going to be used to make a part, all the tools used are set up in this fashion. Usually, each tool will be configured so that zero is where the tool just contacts the top of the starting material.

Since the depth of material cut may vary from run to run, it may be necessary to do this procedure frequently. A height setting device (herein referred to as a height setter) can be helpful in doing this. Essentially this is an object of a precisely known height (for instance, 1 inch) that is used as an indirect reference point for setting The Z-axis height. The following is a step-by-step account of how to use one:

1. Place the height setter on the material to be cut (the stock material).
2. Adjust the height of the Z-axis carefully until the bottom tip of the tool just contacts the height setter.
3. Move the height setter a little bit side to side to determine how hard the tool is pressing into it, and carefully re adjust the Z-axis height as necessary so that the tool is barely touching the height setter.
4. Set the Z-axis coordinate to whatever the height of the device is in the CNC controller software (if the device is 1 inch tall, then set the Z-axis height to 1.00 inch in the software).
5. Remove the height setter.

At this point, the bottom of the tool should just contact the top of the stock material when the Z-axis is repositioned to Z = 0.

It is also possible to directly set the Z-axis height by very carefully adjusting the Z-axis height until the tip of the tool touches the top of the stock material. The use of the tool height setter is not as complex as it may sound from the steps listed above though, and this has some advantages over this direct method. First, it eliminates the possibility of pushing the tool into the work piece and damaging the starting material. Further, the above approach a gives a tactile indication of how hard the bit is actually pushing into the stock material, and will usually give a more accurate result.

Height setters may also be created in a way that protects the cutting tool, such as being spring loaded. It is possible to purchase probes manufactured for this purpose. Unfortunately, they are fairly sophisticated devices which are intended for use with commercial machines. This means that they may be expensive, and may not work without modification on a smaller scale machine.

Cut Order

It is important to consider cut order when creating the G-code for a part. Cut order refers to the sequence in which particular features in a part are cut. Take, for instance, the part shown in Figure 64, a faceplate for a speaker cabinet. The final part is essentially a wooden rectangle with a couple of holes cut in it (labeled as 'a'). It is just as easy to write G-code that tells the machine to cut the holes first as it is to have it cut the outer rectangle first.

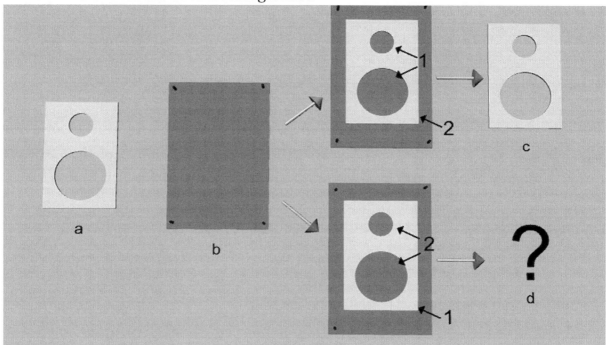

Figure 64: *Cut Order: The numbers indicate the order in which a particular feature of the part (in this case the face for a speaker cabinet) is cut. There is little or no possibility of success if the outer rectangle is cut first, because the material will not be held securely when the interior holes are cut.*

The piece of stock material (labeled 'b' in Figure 64), is a piece of wood that is fixed in position at its corners. Something bad will probably happen if the exterior rectangle is cut before the interior holes, as indicated by the giant question mark (d), because the part will no longer be securely held when these final cuts are made. Cutting the holes first will work much better, because the material is still firmly held when the final cut (the rectangular outer cut) is made. This is a simple example. As parts become more complex, more thought is required to set things up correctly and make sure they are held and cut properly.

Precision, Accuracy, and Reproducibility

Precision, accuracy, and reproducibility are important concepts in machining. Accuracy refers to how close to a true value you are. Precision refers to how tight your tolerances are, i.e. how much variability there is in what is being done. Reproducibility refers to how reliably the machine can repeat some action.

It is possible to be precise without being accurate. For instance, if you saw a robbery and are being interviewed by the police and you gave an incredibly detailed description of the robber's clothes, red hair, and the exact color of four door sedan that was the getaway vehicle, you are being very precise. However, if the robber was actually driving a van and had blonde hair, you are not being accurate.

With a CNC machine, accuracy is how close to a true value you are getting during an operation. If a machine is instructed to cut a square that is two inches per side, but the square ends up being 1.9 inches per side, then the machine has accuracy problems. If it can cut that square to one ten thousandth of an inch, then it is extremely precise. To work properly, you must be both accurate and precise enough for the task for which it is used. A machine used to cut wood to make furniture, for instance, can be less precise than a machine that is used to mill out the cylinders in an engine block. They both need to be accurate, though.

Reproducibility refers to the ability of a machine to make the same move over and over. A CNC machine that has a lot of backlash (i.e. slop between a nut and screw driving an axis) will not be able to reliably reproduce a particular move. To get good results, a machine must have good reproducibility.

When a CNC machine is first up and running, its precision, accuracy and reproducibility should be characterized. This may be done by trying to cut basic shapes such as a circle or square and seeing if they are of the proper dimension, the angles are correct, they are of uniform diameter, and so forth. Making a long move away from and back to a particular spot may be used to characterize reproducibility. Measuring devices such as a dial gauge, micrometer, caliper, and steel square can come in very handy when performing these operations.

Edge Finding

It may also be useful to purchase an edge finding device to accurately determine the coordinates of a work piece (an edge, corner, or center of a hole). There are a number of different types of these. One type spins in the spindle like a cutting tool does, and is moved towards the edge of interest. When it is aligned correctly with the edge, the tip (which can move relative to the body of the edge finder) will come into alignment with the body.

Fancy pressure-sensitive probes are used with professional machines, and light up when an edge is contacted. Automated edge finding has been developed, which relies on the CNC controller software to gradually move a sensor or probe toward the edge of a work piece. Once the edge is contacted by the sensor, a signal is sent to the controller program to indicate the edge location has been found.

Miscellaneous Terms and Topics

Roughing and Finishing Passes

The term 'roughing pass' refers to the use of relatively fast, deep cuts to remove the bulk of the material in a machining operation. The term 'finishing pass' refers to the final pass (or passes) used to create the final, high quality machined surface. Commonly, the roughing pass uses a large cutting bit at relatively high speeds to get rid of large amount of material quickly, and the finishing pass uses a smaller tool at slower speeds to generate the smoother, more detailed final cut.

Jogging

To jog a machine is to manually move the machine into position using whatever directional controls are available. On a professional machine this may be accomplished through a *jog wheel* which may be spun to conveniently move the spindle along a particular axis or by use of a *pendant*, which is a hand held device with buttons or other input devices to allow easy control of machine position. With hobbyist machines, this is typically done using keys on the controlling PC's keyboard. Other computer input devices, such as the Griffin Powermate USB knob or ShuttlePro controller have been suggested for use as alternatives to basic keyboard control.

Fixturing

Fixturing refers to the use of various devices to properly hold material for cutting on a machine. A variety of clamps and fasteners are used to hold a piece (or pieces) of stock securely for cutting. Specific fixtures (jigs) often must be created when there isn't an off-the-shelf option that will work. Fixturing is an art unto itself, and it can be quite tricky to hold a complex part for all the different machining operations to create it. As a part becomes more sophisticated, more time will be spent figuring out how to hold it properly and sequence the cuts during the machining operations.

Rapids

The term *rapid* or *rapids* refers to how fast a positioning move (or moves) can be made, which is typically near or at the top speed of a machine. The sentence "That machine has rapids of 150 inches per minute" means that the fastest rate of movement is about 150 inches a minute, and this is usually the top speed it achieves when repositioning.

Appendix A
Types of Machining

There are different types of CNC machining, and this Appendix lists several of them. They are distinguished from each other for different reasons, which are detailed in the following descriptions. No one type of machining is ideal in all situations; some approaches will be far more appropriate for creating a certain part than others.

2½ axis (2½ D) Machining

2½ D machining typically involves the cutting of outlines, pockets, and profiles to create a given part. Moves in the X and Y directions are often made together (such as when cutting a diagonal line or a circle), but are to a large degree separate of moves made in the Z direction. 2½ D machining is most efficient for creating parts that do not contain complex curved surfaces such as the surface of a sphere.

Both 2½ D and 3 D machining may be done on the same type of machine (a CNC mill or router table for instance). The difference is primarily in how the CAD and CAM programs are used to generate the G-code.

3 axis (3 D) Machining

In 3 D machining, moves in the X, Y, and Z directions are often made simultaneously. Three axis machining is useful when creating a part that contains complex curves. An example might be cutting a propeller shape or hull shape out of wood or plastic. This type of machining isn't efficient for cutting shapes like a simple rectangle or circle; this is more easily accomplished with 2½ D machining.

The toolpaths in 3 axis machining are often created as a grid of parallel lines along either the X or the Y-axis (or both) in which the cutting depth (the Z-axis position) is varied to follow the contours of the part. Figure 51 shows a screenshot from a 3 D CAM program which graphically shows these toolpaths.

Another type of toolpath that is available in 3 D machining (and also shown in Figure 51) is referred to as a 'waterline'. What this means is that instead of cutting a series of lines along the X or Y axes, layers are cut away, starting from the top of the part, moving incrementally down the Z-axis, from the top to the bottom of the stock material. Material is left where the part is supposed to be and removed everywhere else, as if water were being drained from around the part.

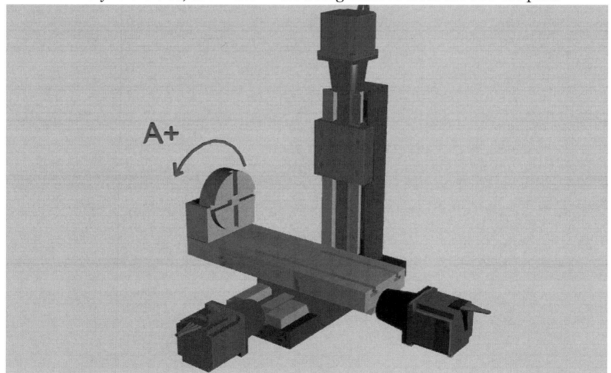

Figure 65: *A rotary table is a common choice for a fourth axis, and is usually labeled as axis 'A'. The arrow indicates the type of rotation that is commonly considered a move in the positive direction. A fifth axis (a second rotary axis) would normally be labeled as the 'B' axis.*

Four Axis Machining

The term 'Four axis' or 'Fourth axis' usually refers to a machine with the typical X, Y, and Z axes, and an additional rotary axis (an axis that rotates instead of moves in a straight line). It is possible to fit a rotary table with a CNC motor and control the angle during a machining run using G-code. Specialized CAM software is necessary to create the G-code. The use of a fourth axis allows things to be cut "in the round" (i.e. from multiple angles) in a single pass, without physically removing and reorienting the work piece.

Use of a fourth axis requires the creation of G-code to control the angle of the rotary axis. Figure 65 illustrates a mini-mill with a rotary table attached oriented along the X-axis (these drawings are based roughly on Taig and Sherline products).

Five Axis Machining

A fifth axis may refer to a second rotary axis mounted perpendicular to the first rotary axis, or it may refer to a pivoting head which enables adjustment of the angle of the cutting tool. This is relatively advanced compared to 2½ and 3 D machining and will require a sophisticated CAM program to generate the G-code.

Alternatives to Drills and End Mills

Most of this book assumes that the user is going to be using a spinning cutting device (drill bit, end mill, etc.) when operating their CNC machine. This chapter discusses some alternatives to cutting as a production method, and what their benefits are. Some of these techniques are more accessible to DIY than others, and each presents a different set of benefits and challenges.

Electrical Discharge Machining

Electrical Discharge Machining, or EDM, uses very high voltage to cut metal. In wire EDM, the cutting tool is essentially a wire which is slowly moved into the material, creating a series of sparks between it and the material to remove the material. This technique is limited to machining conductive material, however, as it relies on arcing from the electrode to the piece being machined. This technique has the benefit that it may be used to machine very hard metals successfully, and is capable of making intricate cuts. As of this writing, DIY plans for a wire EDM machine are available from at least one of the sources listed in the reference section. This machining technique can be relatively messy (usually there is a fluid bath involved), produces fumes, and may represent a fire hazard due to the production of sparks.

Plasma Cutter

A plasma cutter is a device designed to make cuts in metal, and may be automated by the use of CNC. It uses a combination of high voltage and gas to produce plasma. Plasma is a conductive state of matter where a gas has become ionized (had some electrons pulled off of the gas molecules). The heat of the plasma is used to melt the metal while the force of the blowing gas is used to remove the metal. Like EDM, plasma cutting can make intricate cuts. This technique requires that the plasma torch be positioned at a consistent height above the metal. Kits for creating a torch height control are available for those interested in building a plasma cutter.

Foam Cutting

The idea in foam cutting is straightforward; use a hot wire of some type to cut (melt through) polystyrene and other types of foam to create glider wings, props, moldings, and other parts. This technique requires the use of a wire which is heated to a specific, controlled temperature. For most hobby uses, foam cutting doesn't require a machine that is heavy duty, as the forces generated moving cutting wire through foam are relatively low. This means that relatively light duty, inexpensive components may be used to build one.

Kits, components, plans, and related information are easily found on the Internet. This technique creates fumes from the melting of foam, so this will need to be considered when using this type of machine. Foam may also be milled as well as cut with a hot wire.

Lasers

A laser is a device designed to produce an intense and coherent (highly aligned and directional), beam of light of a specific frequency (wavelength). High power lasers can generate intense amounts of energy and put it in a very small area, making them useful as a cutting tool. Lasers may cut a variety of materials, including metals and plastic. They can make extremely fine cuts that generate little waste, and can do things that are hard to do with traditional cutting methods, such as make very intricate patterns, and sharp angle cuts. Simple implementations of laser cutting are generally restricted to cutting relatively flat materials, such as sheet goods, or for engraving.

Development of a CNC machine that uses a laser requires care, especially with high power lasers. A laser can damage someone's eyesight in a fraction of a second. Some lasers, such as CO_2 lasers, emit radiation at wavelengths invisible to the human eye, heightening the possibility of error and injury. Anyone considering using a laser in a DIY CNC device should learn how to safely operate one before proceeding.

Water Jet Cutting

Water jet cutting uses a powerful stream of water containing abrasive material to erode the material being machined. An advantage of this is that a narrow cut width can be produced, allowing for the creation of relatively intricate cuts. Another advantage is that it is essentially a cold cutting process; it produces little heat. This is advantageous when heat may damage or weaken the part being produced. Water jets require a very expensive pump that is capable of generating extremely high pressures and can also be dangerous.

Additive Techniques

The previously discussed CNC techniques focused on the removal of material from a starting block of material through cutting, eroding, melting and so forth. There are CNC devices that create parts by an additive process, where a part is made by progressively building it up from some starting material.

Stereolithography

Stereolithography machines are computer controlled devices that create parts layer by layer. They typically use a pool of photopolymer and a UV light source which is scanned over the polymer to activate it at various locations, causing the photopolymer to solidify (polymerize) in specific locations. The part is built incrementally as each layer of the part is solidified. This technique can produce shapes that are extremely difficult to create by other means. I am not aware of DIY versions of this type of machine, but it is an interesting technology that might be worth studying.

There is a more basic CNC device that works similarly to the stereolithography process described above. A Cornell graduate student has created a 'fabber' that is relatively simple and inexpensive compared to commercial stereolithography machines. A platform moves in the Z dimension, and a syringe of material is positioned in the X and Y dimension by stepper motors with integrated lead screws. The material in the syringe is dispensed in small amounts and a part is built up layer-by-layer. This design incorporates a four axis stepper driver to control the X, Y, and Z axes, as well as one to control the syringe.

It may be worthwhile to examine a picture of this machine because it is constructed of clear acrylic, and most of the linear guides, motors, electronics, and other parts are easily seen in a single view (see Appendix B for a reference). This gives an overview of the essential parts of a 3-axis CNC machine, even though they may look a lot different.

Other Applications

The following applications fall into a miscellaneous category.

Circuit Board Milling

Many CNC enthusiasts have used CNC for the creation of circuit boards for prototyping and limited production purposes. The approach is to take a copper clad board (a circuit board that has a layer of copper on at least one side), and use a CNC controlled mill or small router table and mill out traces and pads for electrical components. A steep-angle engraving bit is used to cut a thin line through the copper surface of the board through to the epoxy fiberglass material beneath to create circuit traces. Additionally, precise drilling may be done by the same CNC machine to make holes for the wire leads of the resistors, capacitors and other parts.

Software is available that will automatically convert files from circuit board design software into G-code. The software determines how to cut appropriate traces from the imported files, and can cut more than one side. It is also possible to use CAD software and map out traces by drawing lines, then export the drawn file as a DXF and convert it into G-code.

Pick and Place

Efforts have been made to create simple pick and place machines using a CNC controlled mini mill. Pick and place machines are robots designed to automatically load electronic components onto a circuit board. They are used for high volume production of circuit boards. The development of a relatively simple low cost one for prototyping and small production runs would be valuable for individuals developing circuits.

One design uses a mini mill and pneumatic valves to pick up parts. The valves are used to open and close vacuum and pressure lines which make up part of the mechanism used to pick up and release small surface mount resistors and capacitors. A CNC mill provides the accurate positioning for repeatedly placing parts over and over.

Solder Paste Dispensing

Yet another circuit board related application is to use a CNC machine to dispense small volumes of solder paste onto circuit boards prior to placing resistors and capacitors. Combined with a homebrew pick and place machine, this would greatly automate small scale circuit board assembly. One example uses a CNC controlled syringe and a CNC mini mill; the plunger is depressed to dispense small amounts of solder paste at pre-programmed spots around a board.

3 D Scanning

3 D models may be automatically 'built' by a scanning process that combines motion control technology and computer software. One of the sources listed in the reference section has software and plans for a 3 D scanner which combines a CNC controlled turntable, digital video camera, laser, and specialized software to make detailed 3 D scans of objects. There are commercial 3 D scanners available, but they tend to be fairly expensive.

Appendix B

Resources

Sources of information, software, parts, and plans (retail and surplus).
There is no specific endorsement implied in this list, it is supplied to help point readers to potentially useful information and parts. I've tried to stick to sources that have been around for a while. For the sake of disclosure, I have worked with the owners of High Tech Systems and Denver CNC on various projects over the past few years. Many of the sources listed may be located easily by typing the name into a search engine.

Surplus Electronics

Name	Internet Address	Description
Electronic Goldmine	http://www.goldmine-elec.com/	Motors and surplus electronics
Apex Jr	http://www.apexjr.com/	Surplus electronics
Ebay	http://www.ebay.com/	Auctions for misc parts
Digikey	http://www.digikey.com/	Electronics parts retailer
U S Digital	http://www.usdigital.com/	Encoders
Phidgets	http://www.phidgets.com/	PC interface products
Winford Engineering	http://www.winfordeng.com/	Breakout and relay boards
Embedded Acquisition Systems	http://www.embeddedtronics.com/	Servo boards & misc. electronics
Homann Designs	http://www.homanndesigns.com/DigiSpeedDeal.html	Speed controller

Sources of CNC Plans

Name	Internet Address	Description
Data Cut Router Plans	http://www.data-cut.com/	CNC router table plans
John Kleinbauer	http://www.crankorgan.com/	Low-cost CNC designs
Bob Campbell designs	http://www.campbelldesigns.com/	Plans and accessories
Solsylva	http://www.solsylva.com/	CNC plans (belt, rack and pinion based)
Machine Tool Camp	http://www.machinetoolcamp.com/	CNC router table plans
Camtronics	http://s120220635.onlinehome.us	EDM & Scanner Plans
Aeropic	http://aeropic.free.fr/indexenglish.htm	A CNC foam cutter design
Compton Systems	comptonsoft.com/cnc/index.php	CNC kits and accessories
HomeShopCNC	http://www.homeshopcnc.com/	CNC parts
Build your CNC	http://buildyourcnc.com/default.aspx	CNC router design
Aeropic	http://aeropic.free.fr/indexenglish.htm	CNC foam cutter design

CNC and Milling Parts and Accessories

Name	Internet Address	Description
Techno Isel	http://www.techno-isel.com/	A wide variety of CNC
High Tech Systems	http://www.hightechsystemsllc.com/	Hobby machining accessories
Little Machine Shop	http://www.littlemachineshop.com/	Hobby machining accessories

Software

Name	Internet Address	Description
ArtSoft	http://www.machsupport.com/	Mach 2,3 and LazyCam
GRZ Software	http://www.grzsoftware.com/	Mesh CAM 3 D CAM
DAK Engineering	http://www.dakeng.com/	Turbo CNC, ACE Converter
Desk Proto	http://www.deskproto.com/	CAM Software
MecSoft	http://www.mecsoft.com/Mec/	Visual Mill CAD/CAM
EMC	http://www.linuxcnc.org/	Free Linux based Controller
Softsquad	http://www.softsquad.com/	G-code editor
Discriminator	http://www.cncedit.com/	G-code editor
Sheet CAM	http://www.sheetcam.com	2½ D CAM program
PCB-Gcode	http://groups.yahoo.com/group/pcb-gcode/	PC Board milling software
CNC Toolkit	http://www.rainnea.com/cnc_toolkit.htm	5 and 4 axis machining
Dolphin CADCAM	http://www.dolphincadcamusa.com/	CAD CAM software

Miscellaneous Parts

Name	Internet Address	Description
Stock Drive Products	http://www.sdp-si.com/	Gears and such
McMaster Carr	http://www.mcmaster.com/	Industrial parts (rack and pinion, belts, rail)
Think and Tinker	http://www.thinktink.com/	Cutting tools
Harbor Freight	http://www.harborfreight.com/	Miscellaneous tools
Nook	http://www.nookindustries.com/	Screws and other CNC devices

CNC Applied to Art

Name	Internet Address	Description
Art of Motion Control	http://www.taomc.com/	CNC as applied to art
Bathsheba Art	http://www.bathsheba.com/downloads/	3 D printing article & art

Informative Web Sites

Name	Internet Address	Description
CNC Zone	http://cnczone.com/	CNC and Machining forums
Desk Top CNC	http://www.desktopcnc.com/	Small mill comparisons
Dust Collection	http://billpentz.com/woodworking/cyclone/Index.cfm	Cyclone design and info about dust and health
Stepper Motors	http://en.wikipedia.org/wiki/Stepper_motors	Wikipedia article on steppers with animations
Wiring Information	http://www.powerstream.com/Wire_Size.htm	Current capacities of wire, etc.
NIST G-code info	http://www.isd.mel.nist.gov/personnel/kramer/pubs/RS274NGC_3.web/RS274NGC_33a.html#999262	Description of G-codes
Desktop CNC	http://www.desktopcnc.com/	Comparisons of mills, etc
Machine Design Article	http://www.machinedesign.com/ASP/strArticleID/56489/strSite/MDSite/viewSelectedArticle.asp	Discussion of lead screws and ball screws
Jones on Steppers	http://www.cs.uiowa.edu/~jones/step/	Stepper motor tutorial
Yahoo group	http://groups.yahoo.com/group/CAD_CAM_EDM_DRO/join	CAD CAM group
Fab at Home	http://www.fabathome.org/wiki/index.php?title=Main_Page	3 D fabrication for the masses
RC Modeling and CNC	http://www.rcscalebuilder.com/Tutorials/cnc/getting_started_in_cnc.htm	An overview of CNC with as applied to RC modeling
CNC project page	http://www.ciciora.com/index.html	Pick and place and solder paste dispenser using a mill
Engineering Fundamentals	http://www.efunda.com/	Basic engineering information

Appendix C

Tools

The following is a list of tools that may prove useful when working on a CNC machine. Some of the most obvious ones have been excluded (ratchet set, screw drivers, and so forth).

Taps and Tap Handles

A tap is used for cutting threads in metal and hard plastic. Taps come in a variety of sizes and thread profiles and are specified as to diameter and thread pitch, similar to the specifications for screws. For instance a tap specified as ¼-20 will cut threads for a ¼" diameter screw with 20 threads per inch (a.k.a. a ¼ - 20 screw). A tap handle is usually required for holding taps.

Machining fluid is helpful when cutting threads in metal with a tap as it reduces friction, making the task easier, and will reduce the chance of snapping an expensive tap.

Cut Off Wheels

Cut off wheels are discs formed from abrasive material, and are available for use with a circular saw. These are invaluable when trying to cut hardened steel such as a round rod to length for use as a linear rail. Hardened steel is extremely difficult to cut by conventional methods (e.g. a hacksaw). These do not produce a clean edge when cutting, though.

Drill Press

A small drill press may be a good investment if one is not already available. It is often necessary to drill holes repeatedly and accurately through metal when constructing a CNC machine or performing a conversion. Trying to do it by hand is difficult at best, and small bench top drill presses are relatively inexpensive.

Cut Off Saw

A cut off saw is designed to make repeated, right angle cuts through stock and when equipped with the right blade (see below) it can make cutting metal easier, faster, and more accurate.

Carbide Tipped Circular Saw Blades

Circular blades with carbide tips are available for use with a cut off saw, and produce nice, clean edges when cutting aluminum. Decent blades can be expensive, however. If you only need a limited number of cuts made, it might be better to have an aluminum supplier do it.

Machining Fluid

Machining fluid is designed to make metal working easier by lubricating cutting surfaces, and is useful when making cuts in metal. There are different types available, some are petroleum products and others are biodegradable. A little lubricant like WD-40 may suffice.

Dial Indicator or Dial Gauge

A dial indicator is a measuring device with a plunger and a clock-like indicator face. These gauges are used for making precision measurements and can measure differences of less than on thousandth of an inch. These devices are useful for checking alignment, roundness, and thickness, among other things, in a variety of situations. Lower end ones (inexpensive imports) are available for approximately $10 from some suppliers.

Caliper

A caliper is another device for making precise measurements, and typically features an adjustable jaw which is moved into position around or inside a part to be measured. Calipers with digital readouts are becoming more common and inexpensive, and provide a convenient zeroing function.

Multimeter

A multimeter (or multitester) is useful for troubleshooting electrical and electronic devices. They are designed to measure resistance, voltage, current, and so forth. Multimeters are handy a variety of tasks when building a CNC machine, such as verifying the correctness of wiring, whether or not a power supply is putting out the correct voltage, and determining which wires are the ends of the same coil in a stepper motor, among others. Multimeters vary greatly in price, and higher priced models offer greater functionality, accuracy, and durability. Relatively inexpensive ones should be sufficient for most of the tasks related to constructing a CNC machine, however.

Countersink

A countersink is a device designed to create a cone shape to allow a bolt or screw to lay flush with the surface of the material it is being installed in to. A countersink is typically used after a hole has been created with a drill bit.

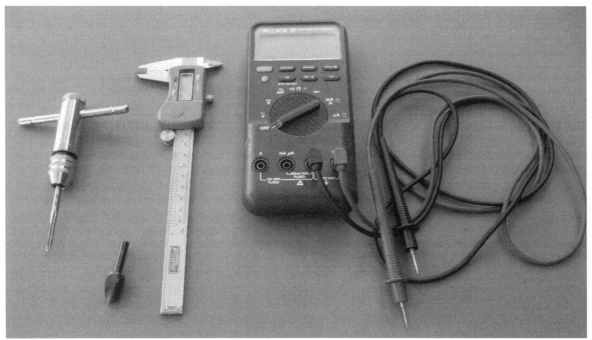

Figure 66: *A selection of the tools described. From left to right, a tap handle and tap, a countersink/deburring tool, a digital caliper, and a multimeter.*

Deburring Tool

A deburring tool is a device used to smooth sharp metal edges and remove unwanted edge material after a cut has been made in metal. Deburring tools come in a variety of styles, but commonly have a plastic handle which holds a swiveling metal blade.

Forstner Bits

Forstner bits are designed to create flat-bottomed holes in various diameters. Although not suitable for machining metal, these bits are good for plastic and wood, and may prove handy in creating certain parts from these materials.

Appendix D

Electricity and Electronics

This is the most rudimentary of introductions, and is largely a collection of definitions which may be useful when dealing with some of the electronics related topics in CNC. Please consult a textbook or other resource for a real treatment of the subject.

Introduction

Electricity, in general terms, refers to the flow and/or presence of electrical charge, and electronics generally refers to circuitry designed do useful tasks with electricity.

Two fundamental characteristics of electrical flow are current and voltage. In a useful (but imperfect) analogy relating to the flow of water, current may be viewed as the amount of water flowing. Voltage may be viewed as the strength of the flow (pressure).

Conductors

Electrons can flow easily through certain class of materials, known as conductors. The most commonly thought of conducting medium is a wire (be it made of copper or some other metal), but many other substances are conductive.

Diode

A diode is a semiconductor device that will act as a conductor for electrons in one direction but not the other.

Semiconductors

Semiconductors are devices that have some ability to conduct, but this ability is highly dependent upon the electrical field placed across the material. This property allows them to act as switches and as amplifiers, and they are used in almost all modern electronic devices. Most semiconductors are manufactured from silicon that has had small amounts of certain elements added to them to tailor their conductive properties.

Alternating Current and Direct Current

Alternating current (AC) is electrical current that reverses direction repeatedly over time through a circuit. Direct current (DC) refers to current that flows in one direction only.

Resistance and Resistors

Resistance is the characteristic of a given material to resist the flow of electrons. For instance, a piece of copper has very little resistance to this flow, but materials like carbon do not allow electricity to flow as easily (and an insulator is almost completely resistant to the flow of electricity). Resistors are electronic components that are designed to have a specific resistance. The fundamental unit of resistance is called the Ohm.

Capacitance

Capacitance refers to the ability of an electronic device to store charge. Capacitors are devices specifically designed to store a quantity of charge and are created by placing two conducting surfaces near each other and separating them with an insulating material called a dielectric. The basic unit of capacitance is the Farad, but most capacitors used in practice are a small fraction of this amount, so a more commonly used unit is the microfarad or nanofarad which are a millionth and a billionth of a farad, respectively. Capacitors have a rated voltage, and when this voltage is exceeded, the dielectric may break down causing the capacitor to fail.

Some capacitors have polarity; for instance, electrolytic (a particular type of capacitor) capacitors are polar. One lead must be hooked up to positive voltage and the other lead must be hooked up to negative voltage. Failure to properly hook up an electrolytic capacitor can result in the explosion of the capacitor when the circuit is powered up. An additional safety issue with large capacitors is that they can hold a dangerous level of charge even after they are removed from a circuit or a circuit is turned off.

Ground

In electrical terms, ground refers to a zero volt reference (also known as a *common*) point in an electrical circuit. The voltage specified in a circuit at a given place (e.g. a point in the circuit that operated at +5V or -12V or whatever) are measured relative to this zero point (0 Volts).

Inductor

Inductors are a basic electrical component created by winding insulated wire (or some other conductor like metal foil) in to tightly wound loops known as coils (coil is another name for inductor). They have the property that they resist changes in voltage of the current running through them. Additionally, they can store energy for short periods of time. Motors use tightly wound coils of wire to create an electromagnetic field, and their performance is characterized to some degree by the inductance of these windings.

Insulator

Insulators are materials that are almost entirely resistant to current flow, such as the plastic material surrounding an insulated wire.

Logic Circuit

A logic circuit is a circuit that performs logical operations (such as binary addition and subtraction). In logic operations, only two states (0 and 1 or off and on) are possible, and are represented by two different voltages.

Logic Level or Logic Level Circuitry

The expression *logic level* refers to the voltages at which logic circuits commonly operate. The voltages that represent off and on (or zeroes and ones) are zero volts and a higher voltage (5 volts for example). In a motor controller box, for instance, the step and direction signals sent by the printer port are based on these two voltage values to indicate an off or on state. A stream of these pulses can be used to transmit data and are used (with other circuitry) to trigger and control CNC devices.

Ohm's Law

Ohm's Law: Expresses the relationship between voltage, resistance, and current (amperage) in a circuit. The relationship may be expressed mathematically as:

$$V = I*R$$

Other variations include I= V/R or R = V/I, where V=voltage, I=current, R=Resistance.

Optoisolator

An optoisolator is a device designed to transmit an electrical signal in a non-conductive way (i.e. the electrical signal is turned into an optical signal and then turned back into an electrical signal so information is still transmitted, but there is no electrically conductive path). These devices are used in some motor drives and breakout boards to protect sensitive circuitry by isolating it from voltage spikes (short bursts of high voltage) that may be created by some of the devices to which they are connected.

H-Bridge

An H-bridge is a circuit designed to allow current to be applied in both directions across a motor, and is used as part of servo motor drive circuitry.

Shielding

Shielding is a conductive material used to limit electrical interference which may cause electronic devices to malfunction.   Many electrical components (like transformers) produce electromagnetic emissions which may cause current (i.e. noise) in nearby circuits. Some transformers have a metal band wrapped around them or are placed in a metal housing to help reduce the noise they produce. Insulated wire is often wrapped in a foil sleeve to keep the wire from picking up electromagnetic noise like an antenna (see Figure 44).

Watt

The watt is a unit of electrical power. For instance, a 100 watt bulb is brighter (converts electricity into light faster) and uses more energy than a 60 watt bulb for a given length of time. Electrical motors may be specified in terms of watts as well. The higher the watt rating for a motor, the greater amount of mechanical work it can do in a given period of time.

Appendix E

Wiring Diagrams

The following wiring diagrams are basic power supply and motor hookup schematics for a simple motor controller box based on either stepper or servo motor drives. Some details, such as how to hook up the motor drives to a breakout board, or specific details for a particular model of motor drive have been left out for simplicity. Make sure to consult the manufacturer's directions for specific installation details for their particular product. Some drives may require additional components, additional power supply voltages, and other add-ons to make things work properly. Additional features (an E-stop, or relays for instance) may require additional voltages, devices, and more complex wiring.

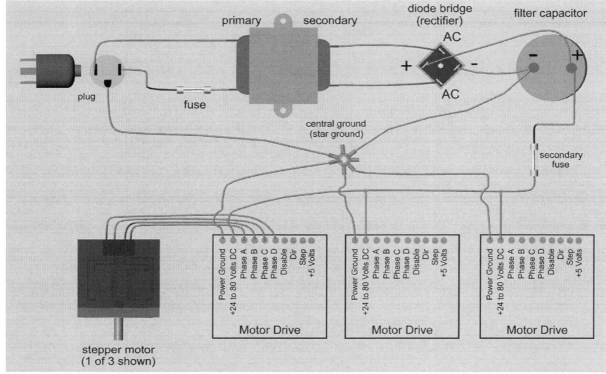

Figure 67: *A basic wiring diagram for a stepper motor box (roughly based on Geckodrive stepper motor drives). This power supply is wired for a simple transformer with a two lead primary and a two lead secondary. A fuse between the primary winding and the wall outlet is an essential safety feature. A central ground point (also known as a star ground) is connected to the ground pin on the plug and to other electrical ground points (i.e. 0 volts) such as the power ground on the motor drives on and the filter capacitor. The star ground is also connected to the case if the case is conductive. This will cause any short circuits to the case to be shunted to the ground pin on the plug.*

Wiring Diagram – Stepper Motor Version

Figure 67 shows a layout for a simple 3 axis controller box, based around Geckodrive stepper motor drives. Different motor drives will vary somewhat in their individual layouts and requirements. Technical reference materials specific to a particular model of drive (either from the manufacturer or from others who have successfully used the product) should be consulted for specific details on installation and use. It is common for transformers to have more than a single pair of leads for the primary and secondary (this was chosen for simplicity).

Additional leads may be present to provide extra secondary voltages, or to provide flexibility in wiring, or to allow the transformer to be wired for use with either a 120 or 240 Volt input voltage (the same transformer may be wired to use either one or the other as needed). Often a transformer will have a diagram either on the transformer itself or with technical documentation that comes with it that shows how to wire it for a particular use.

It is absolutely essential to include a fuse of in line with the primary of the power supply transformer to limit the maximum current the controller box may draw. If there is a short circuit or a motor driver fails and draws excessive current, this fuse will burn up and stop current flow, preventing a dangerous over-current situation. Additional fuses in line after the rectifier filter capacitor may or may not be necessary. If a drive for some reason shorts out and it doesn't cause the primary fuse to blow, the secondary fuse (of a smaller rating) may blow and cause all the drives to lose power. This may protect the machine or keep the cutting tool from being damaged.

Each motor drive requires a separate parallel port pin for both the step and direction signals. These particular drives need a +5V source for the motor drive circuitry. Details regarding wiring to the parallel port are covered in Chapter 8. The drives in this diagram are for *bipolar* stepper motors. The drives have the following hookups (some terminals have been excluded because they are specific to this brand of motor drive):

Power Ground..................... Motor Power Supply Ground
Motor + Voltage .... Motor Power Supply + Voltage
Phase A .... One of the leads from a motor winding
Phase B .... The other end of the winding hooked to A
Phase C .... One lead for the other motor winding
Phase D .... The other end of the second winding
Disable .... Will stop motors when shorted to GND
Dir .... The direction signal input
Step .... The step signal input
+5V .... +5V supply from source of Step and Dir signals

These inputs provide a power supply for the motors, hookups for the coils of the bipolar motor, a mechanism for disabling the motors (for use with the limit switches for instance), and step and direction signals.

Differences with a Unipolar Drive

A unipolar motor drive will require an additional lead running from the stepper motor drive to the center tap on the stepper motor. Drives are specific as to the type of motor (unipolar or bipolar) they will run, so it shouldn't be assumed any given drive will work with any given stepper motor.

Wiring Diagram – Servo Motor Version

A servo motor drive for a brushed DC motor will have several leads connected to the encoder, and will have fewer for the motor itself when compared to stepper motors. For the wiring diagram in Figure 68 (based on Geckodrive servo drives) four leads are wired from the encoder to the motor drive and two are wired to the motor. Figure 69 shows a small servo motor on the right; most of the wires are for the encoder, only two actually go to the motor itself.

Power Ground..............	Motor Power Supply Ground
+18 to 80 V DC	Motor Power Supply + Voltage
Arm-	Wired to one of the motor leads
Arm+...........................	Wired to the other motor lead
Err/Res.........................	An error indicator and reset mechanism
Enc+	The encoder + Voltage supply
Enc-	The encoder ground
Phase A	The encoder's A channel
Phase B	The encoder's B channel
Dir	The direction signal input
Step	The step signal input
Common	Ground for the step and direction signals (e.g. a ground pin on the printer port)

As with the stepper motor diagram (Figure 67), there must be a fuse between the outlet and the primary on the transformer. With servo motors, it is probably more important to have a secondary fuse (or fuses) than with stepper motors, as servo motors will readily draw large amounts of current in a stall or heavy load situations. The advantage of the multiple fuses is that it is immediately apparent which drive and motor are drawing excessive current. The disadvantage is that it is still possible for the other motor drives and motors to continue to operate when one has blown its fuse and stopped operating. In some situations this may cause damage to the machine or cutting tool, and may cause the machine to continue on and cut a part improperly.

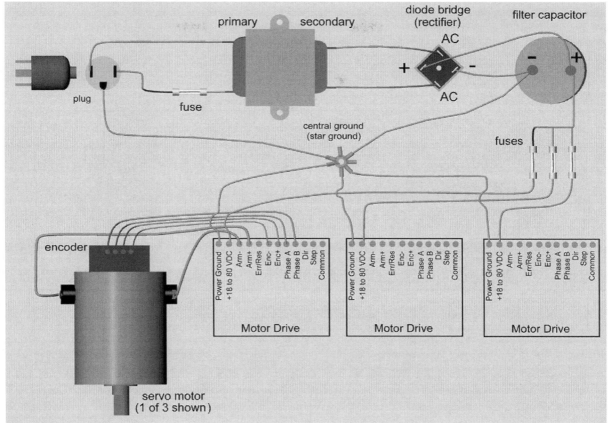

Figure 68: *A DC servo controller box wiring diagram. The power supply hookups are similar to that for stepper motors, but the wiring to the motor (and encoder) is different. The motor itself has two leads, and four leads are used for connecting the encoder. The secondary fuses running to each drive can be used to indicate if a particular drive is drawing excessive current (from an electrical fault or a stall situation). A single fuse could be used instead, as in Figure 67, and would cut power to all drives in an over-current situation where one drive has failed.*

Different motor drives will vary somewhat from what is described here, and the manufacturers should provide detailed instruction on how to configure their products.

Different Ways to Wire a Stepper Motor

The following text refers frequently to figure 70:

It is possible to wire some stepper motors in more than one way. Although four and five wire stepper motors may only be wired in one configuration (**a** and **b**), six wire (configurations **c, d,** and **e**) and eight wire motors (**f, g**, **h**, and **i**) may be wired in different ways. For instance, a six wire motor may be wired to work with a bipolar motor drive by not using two of the leads.

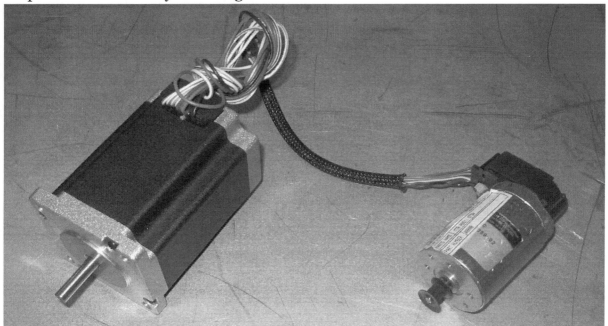

Figure 69: *The eight wire stepper motor at left can be wired in many different ways; unipolar, bipolar, and for different operating characteristics. This flexibility is nice, but adds some complexity during installation. Most of the complexity in hooking up a DC servo motor like the one at the right is with the multiple leads required by the encoder.*

An eight wire motor may be wired in several ways, including for use with either a bipolar motor drive or a unipolar motor drive. Tailoring of motor performance is possible with the different wiring configurations. For instance, wiring in series will produce higher inductance, and may provide better low speed performance. Wiring in parallel will in general provide better high speed performance. Depending on the application, this wiring flexibility with six and eight wire stepper motors may come in handy. The table that follows Figure 70 lists the characteristics of each configuration, and what motor type it may be used with.

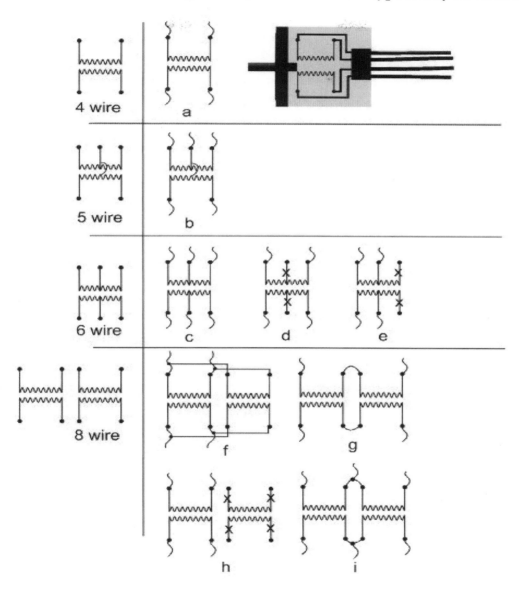

Figure 70: *Stepper motor wiring chart: The X indicates that a particular lead is not being used, and the wavy leads indicate a lead is connected. A description of these different wiring configurations is described on the following table. Six and eight way motors may be wired in a number of different ways, such as for*

unipolar or bipolar use, or for lower or higher inductance. See the chart below for a description.

Description of Different Configurations Shown in Figure 70

	Motor Type	Bipolar or Unipolar	Comments
a	Four wire	Bipolar	Only one possible configuration.
b	Five wire	Unipolar	Only one possible configuration.
c	Six wire	Unipolar	Both center taps wired together.
d	Six wire (four of six wires used)	Bipolar	Higher inductance version than in (e).
e	Six wire (four of six wires used)	Bipolar	Lower inductance than (d) or (h).
f	Eight wire (parallel wiring)	Bipolar	Lower inductance than in (g).
g	Eight wire (series wiring)	Bipolar	Higher than in all other bipolar configurations.
h	Eight wire (half winding)	Bipolar	Lower inductance than all other bipolar configurations.
i	Eight wire (center taps)	Unipolar	Only way to use an eight wire motor with a unipolar motor drive.

Appendix F

Glossary

Miscellaneous Terms

These terms are ones that didn't fit particularly well into any of the discussions of the book, but may be encountered when working in CNC.

Acme Screw

An acme screw is a screw with a specific thread profile. The term acme refers to the thread profile and not necessarily to a brand name. Acme screws are used for linear motion purposes and in heavy duty vises and other applications where a relatively strong, good quality screw is needed.

Climb Milling

Refers to a milling technique where the material is fed in a direction away from the rotation of the cutting tool.

Closed Loop

The phrase 'closed loop' refers to a CNC motor system that uses feedback in its operation. All servo motor systems operate closed loop, because they require feedback (provided by an encoder or some other device) to operate. Most stepper motors are run open loop because steppers do not require feedback to operate.

Conventional Milling

Refers to milling where the material is fed against the rotation of the cutting tool.

Coolant

Coolant is a fluid sprayed continuously on a cutter for the sake of reducing heat build-up during machining operations. It provides cooling by both lubricating the cutting surfaces and by physically absorbing the heat that is generated during cutting.

DRO (Digital Read Out)

A DRO is a digital display which shows the current position of the cutting tool.

Gantry

A gantry is essentially a moving bridge or frame that travels along the X (or long axis) in a CNC router machine.

Hall Effect Sensor

A sensing device used in brushless motor to provide feedback about the rotor's position.

Keyway

A keyway is a slot cut in a motor shaft that integrates with whatever the motor is driving. A keyway provides a mechanism to ensure that the shaft will not slip during operation.

Lost Steps

It is possible for a stepper motor to fail to advance a step or steps when it is supposed to during operation because it can't move the load it is trying to move, because of an electronics problem, or for some other reason. If this happens, it is said to have lost a step or lost steps. Once steps have been lost during a run, future moves made during that run will be offset from where they are supposed to be.

Open Loop

'Open loop' refers to a CNC motor system that operates without the use of feedback. Most stepper motor systems used by hobbyists operate open loop because stepper motors do not require the use of feedback.

Peck Cycle

This is a technique where a drill bit is moved down into and removed from the material being drilled to allow chips formed during drilling of the hole to be cleared. The bit is moved progressively deeper each time until it reaches the desired depth.

Pendant

A handheld device with buttons used to provide convenient control of position on a CNC machine.

Roughing Pass

A quick, first pass in a machining operation made to rapidly remove large amounts of material. This produces a 'rough cut' of a given part.

Swarf

This is the collection of metal chips produced when machining.

Tensioning

Tensioning refers to the process of making a belt appropriately tight between the pulleys that it connects.

Index

Made in the USA
San Bernardino, CA
30 March 2013